U0174547

新理念专业英语（自动化类）

主编　邢金凤　张　洋
副主编　赵　欣　王　琪　田　军　梁　亮
参编　张英平　钱海月　高艳春　李劲涛
　　　刘　爽　梁玉文　冯志鹏　高　岩
　　　周立波

机 械 工 业 出 版 社

本教材遵循"实用为主、够用为度"的教学理念,从职场交际口语开始,然后开展阅读、应用文写作和拓展知识等相关内容。本教材共有 8 个单元,主要内容包括 Introduction、Greeting、Phone Call、Invitation、Reservation、Appointment、Suggestions、Interview。每个单元均通过口语、阅读、写作和知识拓展来提高学生的专业英语语言技能和职场交际能力。

本书可作为高职高专院校自动化类专业的教材,也可供相关专业的英语爱好者使用。

图书在版编目(CIP)数据

新理念专业英语:自动化类 / 邢金凤,张洋主编.
—北京:机械工业出版社,2019.12(2024.8 重印)
ISBN 978 - 7 - 111 - 64185 - 8

Ⅰ.①新… Ⅱ.①邢…②张… Ⅲ.①自动化-英语
-教材 Ⅳ.①TP1

中国版本图书馆 CIP 数据核字(2019)第 263461 号

机械工业出版社(北京市百万庄大街 22 号 邮政编码 100037)
策划编辑:侯宪国 责任编辑:侯宪国
责任校对:张 力 封面设计:张 静
责任印制:张 博
北京建宏印刷有限公司印刷

2024 年 8 月第 1 版第 4 次印刷
184mm×260mm · 9.5 印张 · 186 千字
标准书号:ISBN 978 - 7 - 111 - 64185 - 8
定价:35.00 元

电话服务 网络服务
客服电话:010 - 88361066 机 工 官 网:www.cmpbook.com
　　　　　010 - 88379833 机 工 官 博:weibo.com/cmp1952
　　　　　010 - 68326294 金 书 网:www.golden-book.com
封底无防伪标均为盗版 机工教育服务网:www.cmpedu.com

前　言

在全球一体化的大背景下，我国高素质、高技能人才不仅需要有扎实的专业技术，还要有一定的专业外语知识。虽然市场上自动化类专业英语教材很多，但随着职业院校教学改革的推进，自动化类专业英语教材内容也应不断更新。为了能满足相关院校对自动化类专业英语教材的新需求，我们编写了此教材。

本教材从新时代高素质、高技能人才培养的总体目标出发，以就业为导向，按照"实用为主、够用为度"的原则，突出专业知识的英语表达和职场应用能力，特别注重培养学生的英语对外交际能力。全书共分为 8 个单元，每个单元都有 4 个部分，即"Let's Talk""Let's Read""Let's Write""Learning More"。"Let's Talk"部分以主题展开口语训练，"Let's Read"部分是阅读与专业相关的文章，"Let's Write"部分是常见应用文写作，"Learning More"部分是相关词汇和语法知识的补充学习。在内容选取上，本教材增加了职场英语话题，突出英语的交际性；增加了应用文写作，突出了实用性。另外，练习中还设置了看图片认识专业英语词汇，帮助学生掌握专业词汇，旨在培养学生专业英语知识和英语技能的综合运用能力。本教材可作为高职高专院校自动化类专业的教材，也可供相关专业的英语爱好者使用。

本教材由多年从事专业英语教学的一线教师和专业带头人共同编写而成。其中，邢金凤、张洋担任本教材的主编，赵欣、王琪、田军、梁亮担任副主编，张英平、钱海月、高艳春、李劲涛、刘爽、梁玉文、冯志鹏、高岩、周立波参与了本教材的编写。

由于编者水平有限，书中不妥之处在所难免，敬请广大读者批评指正。

编　者

CONTENTS

Unit One
Introduction

Part One   Let's Talk

Task 1   Read the sentences aloud and do exercises.

■ May I introduce you to...? 请允许我把您介绍给……。

e. g. May I introduce you to Ms. Wang? 请允许我把您介绍给王女士。

■ I'd like you to meet... 我想让你见一见……。

e. g. I'd like you to meet Miss Lewis, our marketing manager.
我想让你见一见我们的市场部经理路易斯女士。

■ I'm..., the... of the company. 我是……，公司的……。

e. g. I'm Li Han, the manager of the company. 我是公司的经理李涵。

■ This is..., he's in charge of... 这是……，他主管……。

e. g. This is Mr. Jacobs, he's in charge of marketing.
这是雅各布先生，他主管营销。

■ Please allow me to introduce... to you. 请允许我向您介绍……。

e. g. Please allow me to introduce Mr. Green to you.
请允许我把您介绍给格林先生。

Exercise 1 Fill in the blanks.

1. 请允许我把您介绍给我们的经理林先生。

   May I introduce _____ to _____, _____?

2. 我想让你见一见我的老板。

   I'd like you to _____.

3. 你好，我是 BBC 公司的秘书琳达。

   Hello, I'm _____, the _____ of the _____.

4. 这是约翰先生，他主管广告。

   This is _____, he's _____ advertisement.

5. 请允许我向您介绍我的经理。

   _____ me to _____ our manager to you.

Exercise 2 Rearrange the sentences to make a dialogue.

A. Hello, I'm Mary. Nice to meet you.

B. Mr. John, may I introduce you to our sales manager, Mr. Black?

C. Nice to meet you, Mr. Green.

D. It's so good to meet you, Mr. Black.

E. And this is Mr. Green, he's our marketing manager.

F. Hello, I'm John. Thank you for picking me up.

_____ _____ _____ _____ _____ _____

Task 2 Role-play

Words bank

sales manager 销售经理	look forward to 希望
cooperation 合作	waiting room 候车室
claim your baggage 提取你的行李	supervisor of the workshop 工厂主管
safety hazards 安全隐患	training 培训
rule 规则	job 工作

Dialogue 1

Mr. Wu　: May I introduce you to Ms. Wang, our sales manager?

Ms. Wang : Welcome to Jilin, Mr. Brown. Our CEO Mr. Zhang Qiang has asked me to pick you up.

Mr. Brown: Nice to meet you, Ms. Wang. I have been looking forward to this trip. I'm glad to visit China.

Ms. Wang : Nice to meet you, too. It's a pleasure to meet a friend here.

Mr. Brown: Me, too.

Ms. Wang : I'm sure we'll have a pleasant cooperation in the future. By the way, how was your trip?

Mr. Brown: Very well. It was very nice all the way.

Ms. Wang : I'm glad to hear that.

Mr. Wu　: Well, let's take a short rest in the waiting room, and then you can claim your baggage.

Mr. Brown: All right.

> **Notes**: Welcome to... 欢迎来到……。
> CEO：Chief Executive Officer 的缩写，译为"首席执行官、执行总裁"。
> by the way：顺便问一下。
> all the way：一直。

Dialogue 2

Jack: Good morning, I'm Jack, the supervisor of the workshop. Nice to meet you.

Bill : Good morning, Jack. I'm Bill. Nice to meet you, too. This is my first time to come to the workshop. Would you show me around? Are there any special things I need to pay attention to?

Jack：Yes, of course. I'd like to point out safety hazards for you. As an employee, you must take an appropriate training so that you can do your job safely. That's quite necessary.

Bill : Oh, safety hazards, what are they? Can you explain more specifically?

Jack：Yes. In fact, it'll be different depending on the workshops you are in. But there is still something in common. For example, getting enough rest and keeping your body physically fit are the basic rules for everyone because these can ensure your full attention to the job.

> **Notes**: show sb. around：带领某人参观　　point out...：指出
> so that：以便　　in common：共同的

Task 3　Find the correct answer for Speaker A

Speaker A	Answer
1. What company are you from, Mr. Wang? _____	a. It's interesting.
2. How do you like the work here? _____	b. No, I didn't. Because I was ill.
3. Which would you prefer, tea or coffee? _____	c. Yes, could you tell me how to get to the airport?
4. Did you attend the meeting last Friday? _____	d. ACC Company. I'm the marketing manager.
5. May I help you? _____	e. Tea, please.

Task 4   Look at the pictures, and choose the correct English expression for each item.

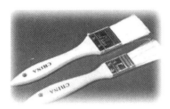

1. 清屑刷（   ）

2. 中心钻（   ）

3. 螺栓（   ）

4. 数显卡尺（   ）

5. 带表卡尺（   ）

A. chip brush   B. bolt   C. digital caliper   D. dial caliper   E. center drill

Part Two   Let's Read

I. Pre-reading Activities.

Task 1   Read aloud the following words and learn the meanings.

workshop ['wɜːkʃɒp] *n.* 车间；工场	production [prə'dʌkʃ(ə)n] *n.* 生产；产品
assembly [ə'sembli] *n.* 装配；集合	operate ['ɒpəreɪt] *v.* 操作；经营
item ['aɪtəm] *n.* 条款；项目	manual ['mænjʊəl] *n.* 手册；指南
machinery [mə'ʃiːn(ə)ri] *n.* 机械；机器	protective [prə'tektɪv] *a.* 防护的
hazard ['hæzəd] *n.* 危险；冒险	symbol ['sɪmb(ə)l] *n.* 标志；符号
explosion [ɪk'spləʊʒ(ə)n] *n.* 爆炸	

Task 2　Write the words according to the English explanation
（Hints：the words from task 1. ）

1. _____ direct or control.
2. _____ a source of danger.
3. _____ a building that contains tools or machinery for making or
repairing things.
4. _____ machines or machine systems collectively.
5. _____ the process of manufacturing or growing something in large
quantities.
6. _____ one of a collection or list of objects.

Task 3　Match the English words with the correct item.

_____ 1. high heels　　A. 口罩
_____ 2. goggles　　　 B. 短裤
_____ 3. gloves　　　　C. 防护鞋
_____ 4. jewelry　　　 D. 高跟鞋
_____ 5. hair net　　　E. 护目镜
_____ 6. mask　　　　 F. 首饰
_____ 7. protective shoes　G. 手套
_____ 8. short pants　　H. 发网

Ⅱ. Reading.

Workshop Safety Rules

Workshop is a room or building where tools and machines are used for making or repairing things, **such as production workshop, assembly workshop**[1]. If you work in a workshop, you need to know some safety rules.

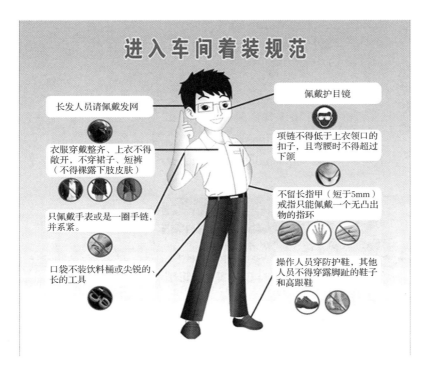

When you operate the machine, it is important to understand the functions and performance of the machine. You need use the operation manual. **There are some safety items described**[2] in the operation manual.

- Anti-vibration gloves must be worn when operating this machine.
- Turn off the machine when it is not in use.
- Don't open this door when the machine is in motion.
- Disconnect power before exchanging or checking electrical motor.

Then, you must wear **protective equipment**[3] while operating the machine. Protective equipment includes the helmet, goggles, work clothes, gloves, face shield, hair net, mask, protective shoes.

To the workers, don't wear jewelry, short pants, skirts, slippers, sandals, high heels in a workshop.

Also, smoking is **a potential safety hazard**[4] for the workshop. It easily leads to fire and even explosion. Therefore, you can see the symbol of "No Smoking" in a workshop.

阅读参考译文

Notes:
1. ...such as production workshop, assembly workshop 意思是"例如，生产车间，组装车间"。
2. there be 句式，表示"有"。过去分词 described 做后置定语修饰名词 safety items。
3. protective equipment：防护装备。
4. a potential safety hazard：一种潜在的安全隐患。

Ⅲ. After-reading Activities.

Task 4 Answer the questions according to the passage.

1. What does workshop mean in this passage?
2. How many safety rules are there in this passage? What are they?
3. What does the protective equipment include?
4. Can the workers wear jewelry, short pants, skirt, slippers, sandals, high heels in a workshop?
5. What's the Chinese meaning of "No Smoking"?

Task 5 Fill in the blanks according to the passage.

Workshop is a room or 1. _____ where tools and machines are used for making or 2. _____ things, such as production workshop, assembly workshop. If you work in a 3. _____ , you need to know some safety rules.

When you 4. _____ the machine, it is important to understand the functions and performance of the machine. You need use the operation manual.

There are some safety items described in the operation 5. _____.

Task 6   Translate the sentences into Chinese.

1. There are some safety items described in the operation manual.
2. Anti-vibration gloves must be worn when operating this machinery.
3. Turn off the machine when it is not in use.
4. Don't open this door when the machine is in motion.
5. Disconnect power before exchanging or checking electrical motor.

Task 7   Complete the following sentences by choosing an appropriate answer from the box.

safety rules   function   turn off   power   symbol   leads to

1. The car has much better _____ than others.
2. Before you leave the conference room, remember to _____ the light.
3. The salesman is performing the main _____ of a new product.
4. What's the special _____ for the workshop?
5. Smoking easily _____ fire and even explosion.

Task 8   Translate the following sentences into English.

1. 你阅读这个产品的操作手册了吗？（operation manual）

_____

2. 操作机器时，你要穿戴保护装备。（machinery）

_____

3. 公共场所禁止吸烟。（smoking）

_____

4. 机器运转中你不能离开。（in motion）

_____

5. 工作人员在检查火灾隐患。（hazard）

_____

 ## Part Three　Let's Write

Notice

通知是用来向大众发布消息的一种方式。通知一般可分为口头通知和书面通知两种。书面通知又分为两种：一种是布告式通知，另一种是书信式通知。

布告式通知类似海报的格式，一个句子可以写成几行，且尽量写在中间，各行第一个字母一般要求大写。语言也尽量简明扼要，通俗易懂。

书信式通知就是以书信的格式书写通知，但是又不完全与书信一致。主要不同之处在于通知要写标题，某些情境下不用写称呼语。

通知的格式：

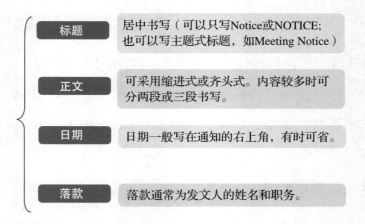

标题	居中书写（可以只写Notice或NOTICE；也可以写主题式标题，如Meeting Notice）
正文	可采用缩进式或齐头式。内容较多时可分两段或三段书写。
日期	日期一般写在通知的右上角，有时可省。
落款	落款通常为发文人的姓名和职务。

Sample 1

> ### Notice
>
> April 24
>
> Dear colleagues,
>
> Please be informed that there is something wrong with our sign-in system. The technical staff has already come to check and they will repair the system within two days. Please come to the office to sign in every morning before 9.
>
> We are sorry for any inconvenience that was caused.
>
> Wang Lihui
>
> Secretary

Sample 2

> ### LECTURE
>
> **Professor Jacob Thomason**
>
> Will give us a lecture on
>
> **Communication**
>
> Time: 6:00 p.m., May 20
>
> Place: Lecture Hall
>
> All are warmly welcome!

Useful Expressions

1. Please be informed that...敬告诸位……

2. I am glad to tell you that...很高兴告知您……

3. The lecture will begin at 3:00 p.m. in Room 101 of Main Teaching Building.
报告地点主教学楼101室，下午3:00开始。

4. Please be on time. 请准时到达。

5. Anyone who is interested in it is welcome to attend the lecture.
欢迎任何感兴趣的人来听报告。

6. All the staff members are expected to attend the meeting.
全体员工务必参加会议。

7. Everyone should bring a pen and a notebook to the meeting.
所有参会人员需带笔和笔记本参会。

8. All the students who are in the first year are welcome to join us.
欢迎所有一年级的学生加入我们。

9. We are very sorry for the inconvenience that was caused.
对由此带来的不便，深表歉意。

10. Everyone is welcome！欢迎大家光临！

Writing Practice

(A.) *Fill in the blanks according to the information.*
说明：请以经理秘书周洋的名义发布一则通知，内容如下：公司因开发新业务将迁至新地址，并为此带来的不便表示歉意。同时借此机会感谢客户多年来的支持，表达期望长期合作的愿望。

Removal Notice

Dear Sir or Madam,

Due to the development of new business, we are going to move to Room 1708 Jiuyang Building No. 35 Siping Street from January 10, 2018.

_____(1)_____（我们的电话号码和传真）are 0431 – 8645708 and 0431 – 8645709 respectively.

We are very sorry for _____(2)_____（为此带来的不便）. We would also like to take this chance to _____(3)_____（感谢您的支持）over the years and hope that we can _____(4)_____（继续合作）in the future.

_____(5)_____

Zhou Yang

Secretary

(1) _____

(2) _____

(3) _____

(4) _____

(5) _____

B. *Write a notice according to the information and design the forms.*

学生会将于本周五（10 月 12 日）举办英语演讲大赛，仅限大一新生参加，演讲题目自拟，演讲时长不超过 5min，请有意者在周四之前到学生会办公室报名。

⟳ Part Four   Learning More

I. Vocabulary

表示"组成"或"构成"的相关词汇

■ **compose** *v.* 组成，构成

（过去式 composed；过去分词 composed；现在分词 composing）

England, Scotland and Wales compose the island of Great Britain.

英格兰、苏格兰和威尔士组成大不列颠岛。

【短语】be composed of 由……组成

Steel is composed of iron and a number of other elements.

钢是由铁和若干种其他元素组成的。

■ **consist** *v.* 组成

【短语】consist of 由……构成

Brass consists of copper and zinc.

黄铜由铜和锌构成。

■ **comprise** *v.* 构成

（过去式 comprised；过去分词 comprised；现在分词 comprising）

These factors comprise your character.

这些因素构成你的性格。

■ **form** *v.* 组成，构成

Four paragraphs form an article.

四个段落构成一篇文章。

■ **be made up of** 由……组成；由……构成

The house is made up of a living room, a bedroom, a kitchen and a bathroom.

这个房子由一个客厅、一个卧室、一个厨房和一间浴室构成。

Practice

(A.) *Read the following sentences and underline the related words that mean* "组成" *or* "构成".

1. This dish is made up of a number of fruits.
2. Sometimes two different types of machines form a complex one.
3. The novel is composed of about six different parts.
4. The drink consists of 75% water and 25% apple juice.
5. These small parts comprise your product.

(B.) *Translate the five sentences above into Chinese.*

1. _____
2. _____
3. _____
4. _____
5. _____

Ⅱ. Grammar

时 态

英语中时态表示动作发生的时间和所处的状态。时态的变化主要是通过动词本身的变化来实现的。

现将英语语法中的 16 种时态总结如下：

	一般	完成	进行	完成进行
现在	一般现在时 do	现在完成时 have done	现在进行时 am/is/are doing	现在完成进行时 have been doing
过去	一般过去时 did	过去完成时 had done	过去进行时 was/were doing	过去完成进行时 had been doing

（续）

	一般	完成	进行	完成进行
将来	一般将来时 will do	将来完成时 will have done	将来进行时 will be doing	将来完成进行时 will have been doing
过去将来	一般过去将来时 would do	过去将来完成时 would have done	过去将来进行时 would be doing	过去将来完成进行时 would have been doing

常用时态中涉及动词的变化如下：

1．一般现在时中当主语是第三人称单数的时候，动词有如下变形：

动词类型	变化规则	动词原形	第三人称单数形式
一般动词	词尾加 s	work like come	works likes comes
以 s，x，ch，sh 结尾的动词	词尾加 es	discuss fix teach finish	discusses fixes teaches finishes
以 o 结尾的动词	词尾加 es	go do	goes does
以辅音字母加 y 结尾的动词	先变 y 为 i，再加 es	fly	flies

2．现在进行时中动词须有如下变形：

动词类型	变化规则	动词原形	现在进行时形式
一般动词	词尾加 ing	work	working
以不发音字母 e 结尾的动词	先去掉 e，再加 ing	take	taking
以重读闭音节结尾的动词，如果末尾只有一个辅音字母	双写末尾辅音字母再加 ing	get	getting

3．一般过去时、完成时中动词的变化如下：

动词类型	变化规则	动词原形	过去式	过去分词
一般动词	词尾加 ed	work	worked	worked
以 e 结尾的动词	词尾加 d	live	lived	lived
以辅音字母加 y 结尾的	先变 y 为 i，再加 ed	study	studied	studied
以重读闭音节结尾的动词，如果末尾只有一个辅音字母	双写末尾辅音字母再加 ed	stop	stopped	stopped
不规则动词	见不规则动词表	dig	dug	dug

Practice

Choose the correct answer.

1. He said his family _____ to Uruguay the next day.

    A. will fly   B. would fly   C. flew   D. had flown

2. The room _____ every day. You can come here anytime you like.

    A. cleans   B. is cleaning   C. has cleaned   D. is cleaned

3. When the police arrived, the man _____ for 15 minutes.

    A. left   B. was leaving   C. had left   D. had been left

4. She _____ fat because of the snack.

    A. was getting   B. is getting   C. will getting   D. had been got

5. —Have you ever been to Thailand?

    —Yes. I _____ there last year with my colleagues.

    A. go   B. went   C. have been   D. was going

6. _____ she _____ on well with her classmates recently?

    A. Is, getting   B. Does, get   C. Is, getting   D. Does, gets

7. I shall tell you what she _____ at one o'clock this afternoon.

    A. has done   B. has been done   C. had been doing   D. was doing

8. Where did you work before you _____ here?

    A. come   B. came   C. are coming   D. will come

9. It's time you _____ a holiday.

    A. had   B. have   C. will have   D. have had

10. His friends _____ for him but he's not in a hurry.

    A. will wait   B. wait   C. have waited   D. are waiting

Unit Two Greeting

Task 1　Read the sentences aloud and do exercises.

■ Long time no see. /Good to see you again.　好久不见。/很高兴再次见到你。

　e. g.　—Hi, Lisa, long time no see. You look very good.

　嗨，丽萨，好久不见。你看起来很不错。

　—Thank you, Mary. It's so good to see you again.

　谢谢，玛丽。很高兴再次见到你。

■ How's everything? /How is it going?　一切还好吗? /过得怎么样?

　e. g.　—Hello, Jack, how's everything?　你好，杰克，一切还好吗?

　—Very good, thank you.　很好，谢谢你。

■ Pretty good. /Everything's fine. /Couldn't be better.

　非常好。/一切都很好。/不能更好了。

　e. g.　—David, good to see you again. How are you recently?

　大卫，再次见到你很高兴。你最近怎么样?

　—Couldn't be better. I have gained the scholarship.

　不能更好了。我获得了奖学金。

■ Same as ever. /Nothing in particular. 还是老样子。/没什么特别的。

　　e. g. —How's everything, Linda? 琳达，一切还好吗？

　　　　—Same as ever. I just feel bored. 还是老样子。我感觉很无聊。

■ Nice to meet you. /Pleased to meet you. 见到你很高兴。

　　e. g. —Nice to meet you, May. Let me show you around the campus.
见到你很高兴，梅。我带你参观一下校园。

　　　　—Thank you very much. 非常感谢。

Exercise 1　Fill in the blanks.

1. 你好，很高兴再次在这里见到你。

　　Hello, it's _____ to _____ you again here.

2. 约翰，很久不见。你过得怎么样？

　　John, _____ time no see. _____ is everything?

3. 我很好。谢谢。你过得怎么样？

　　_____ good. Thank you. _____ about you?

4. 没什么特别的。我想让你见一个新朋友。

　　Nothing in _____. I'd like you to _____ a new friend.

5. 你好，很高兴见到你。

　　How do you do? _____ to _____ you.

Exercise 2　Rearrange the sentences to make a dialogue.

A. Hello, Jim.

B. Pretty good, actually. I've won the computer contest.

C. Same as ever.

D. Hello, Tom. Long time no see. How are you doing recently?

E. Thank you. How about you? How is everything?

F. Congratulations!

_____　_____　_____　_____　_____　_____

Task 2   Role-play.

> **Words bank**
>
> software engineer 软件工程师   mind 介意   exchange 交换
>
> business card 商务名片   lathe 车床   machinery 机械
>
> liquid 液体   exit 出口   emergency 紧急

Dialogue 1

Lisa   : Hi, Rick. Long time no see. How have you been?

Rick   : Pretty good. Thank you. And you?

Lisa   : Same as ever. I want to introduce you to our software engineer, Peter Green.

Rick   : Nice to meet you, Mr. Green.

Mr. Green : Nice to meet you, too. Would you mind exchanging your business card with me?

Rick   : Of course not. Here you are.

Mr. Green : Thank you. This is mine.

> **Notes**：Would you mind...? 你介意……吗? 后多跟现在分词即 doing。
>
> business card：商务名片，一般包括姓名、职务、公司名称、地址、联系方式等信息。

Dialogue 2

Alan：So there are rules to follow in operating the machine. What are they?

Carl：Before operating the lathe, you must read the operating instructions carefully. You must keep in mind some of the key points. For example, you are not expected to adjust, clean, or oil the moving machinery. It is important to shut down your machine before cleaning or repairing. Never leave any spilled liquid, oil or grease on the machine. Clean them up immediately. All warning signs, signals and alarms should be obeyed.

Alan: Oh. There are too many rules for me to remember!

Carl: Yes, there are more rules you need to know. You should keep your work location clean and orderly. Other objects, such as raw materials and boxes should be kept out of the center of aisles. It's quite important to make sure that aisles and walkways, exits and fire doors, are kept clear all the time. That is really of great importance, especially in case of an emergency.

Alan: Got it. Thank you.

Carl: If there's any danger in workplace condition, report it to your supervisor immediately. Never repair the machine by yourself. After finishing your work, remember to shut down the lathe and clear up your work desk.

Alan: All right, I will keep all those in mind.

> **Notes**: keep... in mind：记住，记在心里
> of great importance：意思是"非常重要"，相当于 very important.

Task 3　Find the correct answer for Speaker A.

Speaker A	Answer
1. May I speak to Mr. Liu? _____	a. What e-mail? I haven't got any e-mail yet.
2. Can you help me to start the machine? _____	b. Oh, he is very nice.
3. Have you read my e-mail? _____	c. Sorry, he isn't in.
4. What do you think of your boss? _____	d. Not yet.
5. Abel, have you finished your report? _____	e. No problem.

Task 4   Look at the pictures, and choose the correct English expression for each item.

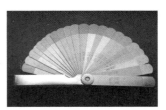

1. 润滑脂（   ）   2. 砂纸（   ）   3. 塞尺（   ）

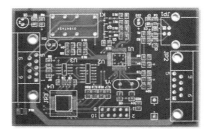

4. 螺母（   ）   5. 电路板（   ）

A. feeler   B. circuit board   C. sandpaper   D. nut   E. grease

◯ Part Two   Let's Read

Ⅰ. Pre-reading Activities.

Task 1   Read aloud the following words and learn the meanings.

minimize [ˈmɪnɪˌmaɪz] v. 把……减至最低数量 [程度]	vibration [vaɪˈbreɪʃ(ə)n] n. 摆动；振动
	danger [ˈdeɪndʒə(r)] n. 危险
cover [ˈkʌvə(r)] n. 覆盖物	clamp [klæmp] v. 夹紧，夹住
prove [pruːv] v. 证明，证实	slow down（使）慢下来
switch [swɪtʃ] n. 开关	execution [ˌeksɪˈkjuːʃ(ə)n] n. 实行，执行
hook [hʊk] n. 钩，铁钩	overhang [ˌəʊvəˈhæŋ] v. 突出，伸出

Task 2　Write the words according to the English explanation （Hints：the words from task 1.）

1. _____ the possibility of being hurt, injured or damaged
2. _____ to reduce or to make sth. less or smaller
3. _____ to show the truth of sth. by evidence or facts
4. _____ to hold sth. tightly to make it fixed or immovable
5. _____ a curled piece of metal to hang things
6. _____ sth. that can be pressed to turn on or turn off a light, machine, etc.

Task 3　Match the English words with the correct item.

_____ 1. CNC machine	A. 防护装备
_____ 2. safety glasses	B. 单块
_____ 3. single-block	C. 安全程序
_____ 4. part	D. 护目镜
_____ 5. safety procedures	E. 进给量
_____ 6. feed rate	F. 锁定功能
_____ 7. lock function	G. 数控机床
_____ 8. protective equipment	H. 零件

Ⅱ. Reading.

Safety Notes for CNC Machine Operations

When operating **CNC machines**[1], there are some preventive actions to follow in order to minimize or avoid any danger.

（1）All the original covers should be kept on the machine **the same as**[2] when leaving the factory.

（2）Wear protective equipment, such as safety glasses, gloves, appropriate clothing and shoes.

（3）Be familiar with the control of the machine before running it.

（4）Make sure to clamp the part in a proper way before running the machine.

（5）It is very important to obey the following safety procedures when proving a program：

● Run the program and check for errors with the lock function of the machine.

● Slow down quick motions with the RAPID OVERRIDE switch.

● Before executing a program，confirm each line with single-block execution.

● Slow down the feed rate of the cutting tool with the FEED OVERRIDE switch to keep from excessive cutting.

（6）Never handle chips by hand and never break long curled chips by chip hooks. If you need to clean the chips，you have to stop the machine first.

（7）Any **overhang tool**[3] should be **as short as possible**[4]，because it may be a source of vibration to break the insert and cause danger.

（8）Make sure the hole-center is large enough to support and clamp the part if you want to support a large part by the center.

Notes：
1. CNC machine：数控机床，其中 CNC 是 Compute Numerical Control 的缩写，machine 译为 "机床"。
2. the same as：与……一致。
3. overhang tool：意思是 "悬挂的工具"。
4. as... as possible：尽可能……，如 as soon as possible，意思是 "尽快"。

阅读参考译文

Ⅲ. After-reading Activities.

Task 4　Answer the questions according to the passage.

1. What should you wear to keep safe before operating a machine?
2. What function can be used to check errors?

3. Which switch can slow down quick motions?

4. What's the function of feed override switch?

5. Can we use chip hooks to break long curled chips?

Task 5 Fill in the blanks according to the passage.

When 1. _____ CNC machines, there are some preventive actions to follow in order to minimize or avoid any 2. _____.

（1）All the original 3. _____ should be kept on the machine the same when leaving the factory.

（2）Wear protective equipment, such as 4. _____, gloves, appropriate clothing and shoes.

（3）Be familiar with the 5. _____ of the machine before running it.

Task 6 Translate the sentences into Chinese.

1. All the original covers should be kept on the machine the same as when leaving the factory.

2. Wear protective equipment, such as safety glasses, gloves, appropriate clothing and shoes.

3. Be familiar with the control of the machine before running it.

4. Make sure to clamp the part in a proper way before running the machine.

5. Never handle chips by hand and never break long curled chips by chip hooks.

Task 7 Complete the following sentences by choosing an appropriate answer from the box.

prove	familiar	minimize	clamp	switch	slow down

1. Turn down the volume _____, the kids are sleeping.

2. My grandparents are very _____ with the neighborhood, they won't get lost.

3. Newton _____ that there is gravity on the earth.

4. You are driving too fast! You must _____ , otherwise you might be in danger.

5. The aim of most factories is to _____ the costs and maximize the profits.

Task 8   Translate the following sentences into English.

1. 运转机器前要先熟悉其操作。(control)

_____

2. 警察用一只手抓紧他的肩膀。(clamp)

_____

3. 程序运行前要先检查是否有错误。(error)

_____

4. 操作机器时应避免任何不恰当的行为。(prevent)

_____

5. 这家工厂生产汽车零件。(part)

_____

Part Three   Let's Write

Memo

Memo 翻译成中文是"备忘录",但是不同于中文的字面意思"备忘",它是一种类似于通知的较为正式的简短便条,并且仅限于在公司或组织内部流通。

备忘录的格式主要分为两个部分，标题部分和主体部分。

1．标题部分：

（1）Memo／MEMO／MEMORANDUM（三种写法均可）

（2）To／TO：（接收人）

（3）From／FROM：（发出者）

（4）Date／DATE：（日期）

（5）Subject／SUBJECT／Re：（主题/事由）

2．主体部分

主体部分就是要传递的信息。因为备忘录主要用于内部沟通交流，所以通常不用写得太正式。有些人习惯在最后署名，但是因为标题部分已经有"From Line"，所以一般情况下不需要署名。

＊格式的范例

<table>
<tr><td colspan="2" align="center">MEMO</td></tr>
<tr><td>To：（接收人）</td></tr>
<tr><td>From：（发信人）</td></tr>
<tr><td>Date：（日期）</td></tr>
<tr><td>Subject：（事由）</td></tr>
<tr><td>Message：（正文）</td></tr>
</table>

＊备忘录格式也不是一成不变的，根据具体情况标题部分的内容要素会有所增减或者顺序有所改变。下面这个格式是大型公司（有多个分公司或分支机构）常用的备忘录格式。

```
                           MEMO
To：（收信人）

Company：（分公司）

Dept：（部门）

Location & Ext：（地址与分机号）

Re：
_____
Message：（正文）
```

Sample 1

```
                    MEMORANDUM

To:      All staff
From:    Peter White, Manager
Date:    December 20, 2018
Subject: Christmas Holiday Season


    The Christmas and New Year's Holiday will begin
from December 23. It will be much appreciated if
your work can be finished before that and your desk
could be cleaned up before you leave.
    Have fun.
```

Sample 2

> **MEMO**
>
> TO: All staff
> FROM: Mr.John Hampton, HR Manager
> DATE: 15th May,2018
> SUBJECT: New appointment
>
> Here is the announcement of the staff transfer in Sales Department. Tina Petal is promoted to the new manager. Jessica is appointed as her assistant. They will be on their duty from next Monday.

Useful Expressions

1. The purpose of this memo is to announce... 本备忘录是通知……

2. I am writing to inform you of... 兹此通知……

3. This memo is about... 本备忘录是关于……

4. Please note that... 请注意……

5. If you have any questions, please call... at... (phone number)
   如有疑问，请致电……

6. Please give me your response as soon as possible. 请尽快回复我。

7. I strongly recommend that... 我们强烈建议……

8. Our members have discussed about this.
   我们的团队成员已探讨过此事。

9. I will check this matter in person. 我会亲自查验此事。

10. Can you please arrange the journey for me?
    能否帮我安排此次行程？

Writing Practice

A. *Write a memo according to the given situation.*

说明：根据所给信息完成以下备忘录。

送达：全体员工。

发自：部门办公室（Department Office）

发件日期：2018 年 7 月 6 日

主要内容：关于年度野餐聚会（annual department picnic），举办时间为 7 月 14 日，提供全部餐饮还有奖品（prize）。请于本周三（7 月 10 日）之前告知是否参加，联系方式：秘书电话 666-8493。

```
                          MEMO
      To：_____

      From：_____

      Date：_____

      Re：_____
    _____

      Message：_____
      _____
      _____
      _____
      _____
      _____
```

B. *Write a memo according to the requirements.*

说明：给你的同事 Mary Lin 写份备忘录。

时间：2018 年 10 月 23 日 。

内容：假定你是王鹏。由于你明天出差去北京一个星期，不能和大家商讨下个月产品促销的事情，请 Mary 通知你最后的促销方案以及所需做的准备。

 # Part Four　Learning More

Ⅰ. Vocabulary

表示"精度"的相关词语

■ **accuracy** *n.* 精确度，准确度

measuring accuracy 测量精度；frequency accuracy 频率准确度；geometric accuracy 几何精度；machining accuracy 加工精度；dimensional accuracy 尺寸精度；location accuracy 定位精度

■ **to an accuracy of...** 达……的精度

Measuring readings can be obtained to an accuracy of one meter.

获得的测量读数可达 1m 的精度。

■ **within the accuracy of** 在……的精度范围内

We can gain the data within the accuracy of 5%.

我们可以在 5% 的精度范围内获得数据。

■ **the accuracy with/to which**... ……的精度

It depends on the accuracy to which the tools are used.

这取决于使用工具的精度。

■ **accurate** *a.* 精确的

accurate price 准确的价格；accurate reading 准确读数

■ **accurate to**... 精度达……；精确到……

This electronic equipment is accurate to microseconds.

这台电子设备精确到微秒。

Practice

(A.) *Read the following sentences and underline the related words that mean "精度".*

1. All readings are accurate to three figures.
2. Measuring accuracy is very important in the experiment.
3. The error is within the accuracy of 1%.
4. Could you tell me the accurate price?
5. These numbers are known to an accuracy of 5%.

(B.) *Translate the five sentences above into Chinese.*

1. _____
2. _____
3. _____
4. _____
5. _____

Ⅱ. Grammar

名词所有格

名词所有格用于表示所有关系。

1. **名词所有格的构成**

a. 大多数单数名词在词尾加 's：

    e. g. my sister's new dress 我姐姐的新裙子

          his mother-in-law's telephone number 他岳母的电话号码

b. 以-（e）s 结尾的名词复数在词尾加 '，否则加 's：

    e. g. all employees' welfare 所有员工的福利

          children's festival 儿童的节日

c. 以-s 结尾的单数名词在词尾可加 ' 或 's：

    e. g. Burns's poem 彭斯的诗

          James' bicycle 詹姆斯的自行车

2. **名词所有格的用法**

a. 可用于表示人的名词，意思是 "某人的"：

    e. g. This is the manager's office. 这是经理办公室。

          St. Paul's Cathedral 圣彼得大教堂

b. 也可用于表示物的名词，如时间、距离、价格等：

    e. g. Today is yesterday's pupil.

          昨日是今日之师。

          It will be half an hour's drive to get there.

          距离那里还有半个小时的车程。

          Jane bought a 99 dollars' worth of dress.

          珍妮买了一条99美元的裙子。

c. 可表示商店、教堂等，也可表示某人的家：

    e. g. I got a Christmas gift for my mother in Macy's.

          我在梅西百货公司给妈妈买了圣诞节礼物。

          I have to go to the barber's to get my hair done.

          我必须去理发店剪剪头发了。

          We are going to have dinner at Aunt Mary's.

          我们要去玛丽姨妈家里吃晚饭。

3. 名词所有格还可以和 of 一起使用：

e. g. a friend of father's 我父亲的一个朋友

books of Austin's 奥斯汀的书

Practice

Fill in the blanks.

1. _____ (Henry) new address

2. _____ (Lucy) favourite movie

3. _____ (James) birthday

4. _____ (bird) eye view

5. the _____ (teacher) office

6. It's only about ten _____ (minute) walk to school from our house.

7. This is the _____ (city) best theme park.

8. Have you read _____ (Keats) poems?

9. Would you like to come to my _____ (sister) for dinner?

10. I think I just heard a _____ (man) voice.

Unit Three
Phone Call

Part One   Let's Talk

Task 1   Read the sentences aloud and do exercises.

■ This is... from... 这是……

 e. g. This is David from BBC. 这是 BBC 公司的大卫。

■ May I speak to..., please? 请问我可以和……通话吗？

 e. g. May I speak to the customer service manager, please?
 我可以和客服经理通话吗？

■ Hang on a moment, please. ∕ Hold on a moment, please. 请稍等。

 e. g. Hang on a moment, please. I'll check if she's in.
 请稍等，我去看看她在不在。

■ Would you like to leave a message? 您需要留言吗？

 e. g. —Hello, is Mary there? 你好，玛丽在吗？

 —No, she's not in right now. Would you like to leave a message?
 不，她现在不在。您需要留言吗？

■ I'll call back later. 我一会再打过来。

    e. g. —Hello, may I speak to Mary? 你好，玛丽在吗？

    —Sorry, she's out. May I take a message?

    对不起，她出去了。需要我捎信吗？

    —No, thank you. I'll call back later. 不，谢谢。我一会再打过来。

Exercise 1　Fill in the blanks.

1. 你好，我是微软公司的汤姆。（电话中）

    Hello, _____ _____ Tom from Microsoft.

2. 请找一下苏珊。（电话中）

    _____ I _____ to Susan, please?

3. 请稍等，我看看她是否在家。（电话中）

    _____ _____ a moment, please. I'll see if she's in.

4. 抱歉她不在。您需要留言吗？（电话中）

    Sorry, she's not in. Would you like to _____?

5. 不了，谢谢。我会再打回来。（电话中）

    No, thank you. I'll _____ her _____.

Exercise 2　Rearrange the sentences to make a dialogue.

A. Hold on a moment, please. I'll see if he's in.

B. All right. Bye bye.

C. Hello, this is Tom. May I speak to Alan?

D. I'm sorry, he's out.

E. Can I take a message?

F. No, thank you. I'll call back later.

_____ _____ _____ _____ _____ _____

Task 2   Role-play.

Words bank

trade 贸易	customer service 客户服务	available 有空的
leave a message 留言	call back 回电话	freezer 冰柜
delivery 递送	sunlight 阳光	heat 热度
ideal 理想的		

Dialogue 1

Monica：Microsoft Corp. Good morning, Monica speaking.

Alan　：Good morning. This is Alan from Hong Xing Trade Company. May I speak to your customer service manager, please?

Monica：Hang on a moment, please. I'll get her for you.
（one moment later...）

Monica：I'm sorry to have kept you waiting, Alan, but I'm afraid she's not in right now.

Alan　：When will she be available?

Monica：I don't know, sir. Would you like to leave a message?

Alan　：No, thanks. I'll call back later. When can I catch her?

Monica：In the afternoon, I think.

Alan　：Fine. Thanks a lot. See you.

Monica：See you.

Notes：Corp. 是 corporation 的缩写形式，译为"公司"。
I'm afraid...：我恐怕……

Dialogue 2

Deliverer : Excuse me, are you Mrs. White?

Mrs. White : Yes, I am.

Deliverer : Good afternoon, Mrs. White. I'm here to send you the freezer you bought in our shop.

Mrs. White : Thank you. Please come in.

Deliverer : Where do you want me to put it?

Mrs. White : Could you put it in the kitchen, by the window?

Deliverer : Sorry, Mrs. White, but that's not an ideal place for a freezer. It is good to keep the freezer from direct sunlight and heat resources.

Mrs. White : I never knew that. Thank you for your advice.

Deliverer : You're welcome. Enjoy your life!

Notes: keep... from...: 使……远离……
heat resources: 热源

Task 3 Find the correct answer for Speaker A.

Speaker A	Answer
1. Thank you for seeing me off at the airport . _____	a. Not too bad.
2. Haven't seen you for ages. How are you doing? _____	b. I'll pay in cash.
3. Will you help me with the design plan? _____	c. My pleasure.
4. It's 30 dollars. How would you like to pay? _____	d. I like it very much. It's small but comfortable.
5. How do you like your new apartment? _____	e. Sure, I'll finish the report first.

Task 4   Look at the pictures, and choose the correct English expression for each item.

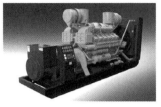

1. 发电机（　　）   2. 电动机（　　）   3. 变压器（　　）

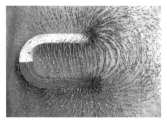

4. 配电盘（　　）   5. 磁场（　　）

A. magnetic field   B. transformer   C. motor   D. switch board   E. generator

Part Two   Let's Read

I . Pre-reading Activities.

Task 1   Read aloud the following words and learn the meanings.

electric [ɪˈlektrɪk] a. 电的；电动的；发电的

field [fiːld] n. 领域，范围

available [əˈveɪləb(ə)l] a. 可获得的；可用的

blower [ˈbləʊə] n. 鼓风机

pump [pʌmp] n. 泵

application [ˌæplɪˈkeɪʃ(ə)n] n. 应用

motor [ˈməʊtə(r)] n. 电动机；发动机

drive [draɪv] n. 驱动器；驱动力

industrial [ɪnˈdʌstrɪəl] a. 工业的，产业的

fan [fæn] n. 风扇

rectifier [ˈrektɪfaɪə] n. 整流器

portable [ˈpɔːtəb(ə)l] a. 手提的，便携式的

Task 2 Write the words according to the English explanation
（Hints：the words from task 1.）

1. _____ a device that uses electricity, gas etc. to produce movement and makes a machine, a vehicle or a boat work.
2. _____ able to be carried or moved easily.
3. _____ connected with electricity.
4. _____ a machine that is used to force liquid, gas or air into or out of sth.
5. _____ the power from an engine that makes the wheels of a vehicle go round.
6. _____ the practical purpose for which a machine, idea, etc. can be used, or a situation where this is used.

Task 3 Match the English words with the correct item.

_____ 1. fan	A. 笼型感应电动机
_____ 2. pump	B. 电动机
_____ 3. motor	C. 交流电动机
_____ 4. AC motor	D. 直流电动机
_____ 5. DC motor	E. 电动机-发电机组
_____ 6. induction motor	F. 泵
_____ 7. cage induction motor	G. 风扇
_____ 8. motor-generator sets	H. 感应电动机

Ⅱ. Reading.

Motors

Each type of motor has its particular field of usefulness. Among all the motors, the induction motor is more widely used for industrial purposes than any other type of AC motor for its simplicity, economy and durability, especially when high-speed **drive**[1] is desired.

If AC power is available, all drives that require constant speed should use cage induction or synchronous motors because of their ruggedness and lower cost. Drives that require varying speeds, such as fans, blowers or pumps, may be driven by wound-rotor induction motors. However, as for machine tools or other machines that require adjustable speed or a wide range of speed control, it will probably be desirable to install DC motors on such machines and supply them from the AC system by motor-generator sets or electronic rectifiers.

The various types of single-phase AC motors and **universal motors**[2] are seldom used in industrial applications, because polyphase AC or DC power is often available. When such motors are used, they are usually built into the equipment by the manufacturers, such as in portable tools, office machinery and other equipment. Generally, these motors are specially designed for the specific machines with which they are used.

Notes: 1. drive 用作名词，可表示驱动力，也可表示起到驱动作用的设备。

2. universal motors 指交直流两用电机。

阅读参考译文

Ⅲ. After-reading Activities.

Task 4   Answer the questions according to the passage.

1. Which type of motors is more widely used for industrial purpose?
2. What are the advantages of induction motors?
3. If AC power is available, which type of electric motors should be used for drives requiring constant speed?
4. Why should all drives requiring constant speed use cage induction or synchronous motors if AC power is available?
5. Why are the various types if single-phase AC motors and universal motors seldom in use in industrial application?

Task 5　Fill in the blanks according to the passage.

Among all the motors, the 1. ＿＿＿＿＿ is more widely used for industrial purposes than any other type of AC motor for its simplicity, economy and durability.

If AC power is available, all drives that require constant speed should use 2. ＿＿＿＿＿because of their ruggedness and lower cost. Drives that require varying speeds, such as fans, blowers or pumps, may be driven by 3. ＿＿＿＿＿.

The various types of single-phase AC motors and universal motors are seldom used in industrial applications, because 4. ＿＿＿＿＿ is often available. Generally, 5. ＿＿＿＿＿ are specially designed for the specific machines with which they are used.

Task 6　Translate the sentences into Chinese.

1. Each type of motor has its particular field of usefulness.
2. The induction motor is more widely used for industrial purposes than any other type of AC motor for its simplicity, economy and durability.
3. If AC power is available, all drives that require constant speed should use cage induction or synchronous motors because of their ruggedness and lower cost.
4. Drives that require varying speeds, such as fans, blowers or pumps, may be driven by wound-rotor induction motors.
5. The various types of single-phase AC motors and universal motors are seldom used in industrial applications, because polyphase AC or DC power is often available.

Task 7　Complete the following sentences by choosing an appropriate answer from the box.

pump	industrial	portable	motor	available	induction

1. There was a little ＿＿＿＿＿ television behind the bar.

2. The machine is used to _____ water.

3. Tickets are _____ from the ticket booth.

4. He started the _____ and left.

5. _____ production has risen by 0. 6% since October.

Task 8   Translate the following sentences into English.

1. 我们客厅里有一台便携式音响设备。（portable）

_____

2. 这个地区有许多就业机会。（available）

_____

3. 大卫是建筑行业里的专家。（field）

_____

4. 我们能从这几口井里抽出干净的水来。（pump）

_____

5. 我开车到机场需要一个多小时。（drive）

_____

Part Three   Let's Write

Note

便条在工作和生活中都应用广泛。它实际上是一种简单的书信，但是书写格式不如书信那样正式。便条的主要目的是尽快地把最新的信息、通知等转告给对方。

便条的语言尽量通俗口语化，简明扼要，直截了当，无须使用客套语言，达到传递信息的目的即可。

便条的基本写作格式：

便条内容和类型不尽相同，可以灵活变通。但各类便条必须包括以下几个基本要素：

（1）Date 便条日期。

（2）Salutation 称呼。

（3）Body 正文。

（4）Signature 署名。

便条的日期通常写在右上角。一般不用写年份，因为便条的内容多半是当日内要办的事情。多数便条由于只是通知或告知一些简单信息，所以在结尾时一般不写结束礼词，只需写上写便条者的姓名。

便条的格式如图：

```
                                          Date
  Salutation
  _____
                        Body
  _____
  _____
  _____
                                       Signature
```

Sample 1

```
                                          May 23

Dear Mr. Miller,

    I had a bad cold yesterday. I went to hospital and the
doctor told me to have a rest in bed. Therefore I can't go to
school today. I'm writing to ask for a sick leave for today
and tomorrow. I shall appreciate it if you'll agree.


                                  Yours respectfully,
                                  Joyce Brown
```

Sample 2

Oct 3

Dear Fiona,

　　I went to your dorm this moming but you were not in. I want to borrow a book *Happy Communication* from you. I will return it back to you immediately after the summer holiday. I wonder if you could contact me anytime in the evening so that I could find appropriate time to visit you.

Julia

Useful Expressions

1. I wonder if you are convenient to call me at 1：00 this afternoon.
   不知您是否方便在下午一点时给我打电话。

2. I am writing to ask for a sick leave for two days.
   我想请两天的病假。

3. Encl.：Doctor's Certificate of Advice. 附：医生证明。

4. Upon receiving this note，please come to my office.
   见到此便条后，请立即来我办公室。

5. Please tidy up your room before I come back.
   请在我回家之前把房间打扫干净。

6. Here is a box of chocolate for your sweet wedding.
   为你甜蜜的婚礼送上一盒巧克力。

7. I hope my absence will not bring you any inconvenience.
希望我的缺席不会给您带来不便。

8. Please favor me with an early reply. 敬请早日回复。

9. I'm grateful if you could give me a reply at your earliest convenience.
如您方便请早日回复，我将非常感激。

10. Please let me know if this is a good timing for you.
请告知我这个时间你是否方便。

Writing Practice

A. Fill in the blanks.

说明：假如你是张小杰，因为感冒且发烧需要卧床休息，不能去上英语课，现在给江老师写一张请假条请假。（随请假条附上医生证明）

时间：10 月 8 日。

_____(1)_____

Dear _____(2)_____,

I _____(3)_____ （非常抱歉）that I shall be unable to attend the English Class due to _____(4)_____ （感冒和发烧）. Enclosed is a certificate from the doctor who said I must _____(5)_____ （卧床休息）for a few days. I will go back to school as soon as I recover.

Yours sincerely,
Zhang Xiaojie

(1) _____

(2) _____

(3) _____

(4) _____

(5) _____

B. *Write a note according to the following requirements.*

　　说明：今天早上威尔森先生（Mr. Wilson）收到一份创维公司（Chuangwei Company）的传真，传真中说创维公司的总经理卡特先生（Mr. Carter）想到生产车间参观并与威尔森先生洽谈新的订单，因此给秘书萨拉（Sara）写下了一条留言。

　　要求：萨拉为卡特先生一行安排好时间和细节。
　　时间：2018 年 8 月 13 日。

Part Four　Learning More

Ⅰ. Vocabulary

表示 "设备" 的相关词语

 facilities *n.* 设备，用具

transportation facilities 运输设备　　sports facilities 体育设备
medical facilities 医疗设备　　supporting facilities 辅助设备
facilities engineer 设备工程师　　facilities location 设备安装位置
facilities planning 设备规划　　harbour facilities 港口设备

■ **apparatus** *n.* 设备，装置，仪器

breathing apparatus 呼吸机　　experimental apparatus 实验装置
beauty apparatus 美容仪器

■ **equipment** *n.* 设备，装备

laboratory equipment 实验室设备   electric equipment 电气设备

common equipment 常用设备   safety equipment 安全设备

electronic equipment 电子设备   office equipment 办公设备

automation equipment 自动化设备

■ **device** *n.* 设备，装置

protection device 保护装置   display device 显示装置

input device 输入设备   mobile device 移动设备

power device 电源装置

■ **plant** *n.* （工业用）大型机械设备

production plant 生产设备

New plant has recently been installed in the factory.

厂里最近安装新的机械设备。

Practice

A. *Read the following sentences and underline the related words that mean "设备".*

1. Different apparatus can be installed in different workshops.

2. There are some scientific devices in the museum.

3. The salesman is selling cooking facilities.

4. Can you introduce your latest research equipment to us?

5. Our company will order a chemical apparatus.

B. *Translate the five sentences above into Chinese.*

1. _____

2. _____

3. _____

4. _____

5. _____

Ⅱ. Grammar

冠　词

冠词是虚词，本身不能单独使用，也没有词义，它用在名词的前面，帮助指明名词的含义。英语中的冠词有三种，一种是定冠词（the Definite Article），另一种是不定冠词（the Indefinite Article），还有一种是零冠词（Zero Article）。

1. 定冠词的用法

定冠词 the 与指示代词 this 和 that 同源，有"这（那）个"之意，但是较弱。定冠词具有确定的意思，用以特指人或事物，表示名词所指的人或事物是同类中的特定的一个，以别于同类中其他的人或事物。

（1）用于指谈话双方都明确所指的人或事物

Give me the pen, please. 请把那支笔给我。

（2）第一次提到的人或物用 "a" 或 "an"，以后再次提到用 "the"。

There is a boy lives next door. The boy goes to school by bike every day. 隔壁住着一个小男孩。他每天骑车去上学。

（3）由普通名词构成的专有名词。

the People's Republic of China 中华人民共和国

the West Lake 西湖　　the Great Wall 长城

the United States 美国　the United Nations 联合国

（4）表示世界上独一无二的事物。

the Sun 太阳　　the Earth 地球　　the Moon 月亮

the sky 天空　　the world 世界

（5）用在表示乐器的名词之前。

He plays the violin. 他会拉小提琴。

（6）用在形容词之前表示一类人或物。

the old 年老者　　the rich 富人

the poor 穷人　　the good 好人

（7）用在方位名词之前。

in the north 在北方　　in the bottom 在底部

on the left 在左边

（8）用在姓氏（复数）之前，表示该姓氏的一家人。

The Woods are having dinner at home.

伍德一家人正在家里吃晚餐。

（9）用在序数词和形容词最高级前。

The manager's office is on the third floor.

经理的办公室在三楼。

Linda is the most beautiful girl in her class.

琳达是她班级里最漂亮的女孩。

2. 不定冠词的用法

不定冠词 a（an）与数词 one 同源，是"一个"的意思。a 用于辅音音素前，一般读作 [ə]，而 an 则用于元音音素前，一般读做 [ən]。

（1）表示"一个"，意为 one。

A knife is a tool for cutting with. 刀是用来切割的工具。

（2）指某人或某物，意为一个确定的事物。

A Wang Ling is waiting for you. 王玲在等你。

（3）代表一类人或物。

Michael is a singer. 迈克尔是一名歌手。

（4）在固定的词组中。

a lot of 很多　　　　　　　many a 许多

in a minute 马上，立刻　　all of a sudden 突然地

after a while 不久　　　　have a try 尝试

in a word 总而言之　　　　as a rule 通常；一般说来

3. 零冠词的情况

（1）表示头衔、职务、职称、身份等的名词前不用冠词。

This is Professor Harvey. 这是哈维教授。

（2）人名、地名、国名等专有名词前通常不用冠词。

Tom went to Australia to see his penfriend.

汤姆去澳大利亚看他的笔友。

（3）书名、标题前一般不用冠词。

Can I borrow *Pride and Prejudice* from you?

你可以借给我《傲慢与偏见》吗?

（4）三餐、球类、学科前不用冠词。

have breakfast/lunch/supper

play football/basketball/soccer /volleyball/tennis

（5）man，woman 表泛指时，一般不用冠词，并且需用单数形式。

Woman wants to be equal with man. 女人想与男人平等。

（6）季节、月份、日期、节日名称前一般不加冠词。

Spring is coming. 春天要来了。

Practice

Fill in the blanks with "the，an，a" where necessary.

1. _____ elephant is much heavier than _____ man.

2. _____ China is _____ old country.

3. The boy will certainly become _____ king of his country.

4. Luna will leave for Shanghai _____ next week.

5. _____ village where I was born has grown into _____ town.

6. _____ old lady in black is _____ university professor.

7. The children all had a good time on _____ Children's Day.

8. My sister had _____ fever, so I had to look after her.

9. My father had to take the medicine twice _____ day.

10. I have seen this movie many _____ time.

Unit Four Invitation

Part One   Let's Talk

Task 1   Read the sentences aloud and do exercises.

- I'd like to invite you to... 我想邀请您去……

  e. g. I'd like to invite you to dinner this evening at Heping Roast Duck Restaurant.

  我想邀请您今晚在和平烤鸭店一起吃晚餐。

- It's very... of you to invite me. 能邀请我，您……

  e. g. It's very kind of you to invite me.

  能邀请我，您真是太好了。

- Shall I pick you up at...? 我……来接您行吗？

  e. g. Shall I pick you up at 5 o'clock then?

  我 5:00 来接您行吗？

- Have you got any plans...? 您……有什么安排吗？

  e. g. Have you got any plans tonight?

  你今晚有什么安排吗？

■ Thanks for your invitation. /Thank you for inviting me to…
谢谢您的邀请。/谢谢您邀请我去……

e. g. Thank you for inviting me to the annual meeting of your company.
谢谢您邀请我参加贵公司年会。

Exercise 1   Fill in the blanks.

1. 我想邀请您参加我的生日宴会。

   I'd like to _____ you to my _____.

2. 能邀请我，您真是太好了。

   It's very _____ of you to _____ me.

3. 我星期六晚上七点来接你行吗？

   Shall I _____ you _____ at 7:00 p. m. on _____?

4. 你下星期日有什么安排吗？

   Have you _____ any plans _____?

5. 谢谢你邀请我参加今晚的家庭聚会。

   Thank you for _____ me to your _____ tonight.

Exercise 2   Rearrange the sentences to make a dialogue.

A. I'd like to invite you to my birthday party on Sunday evening.

B. At 6:00 p. m. I'll pick you up at 5:30.

C. No, nothing.

D. Hello, Susan. Have you got any plans on Sunday?

E. That's great. What time?

F. All right. Thank you. See you then.

_____ _____ _____ _____ _____ _____

Task 2   Role-play.

Words bank

dinner 晚餐   roast 烘烤的   delighted 高兴的
output 产量   process 过程   helmet 头盔
put on 穿上   protect 保护   automate 使自动化
percentage 百分比

Dialogue 1

Gary：Hello, Dave. You are leaving Beijing tomorrow. For business or pleasure?

Dave：Business. I will visit some factories in Changchun.

Gary：Great. Would you like to have dinner this evening at Qing Roast Duck Restaurant?

Dave：Oh, thank you for inviting me. I'd be delighted to go.

Gary：All right. Shall I pick you up at 5:00 p. m. then?

Dave：It's very nice of you.

Gary：And we'll have a dance party after the dinner. There will be some other friends whom you will be glad to meet, I think.

Dave：That's wonderful.

Notes：For business or pleasure? 是习惯说法，译为"公务出差还是旅游?"。
Qing Roast Duck Restaurant：清烤鸭店。

Dialogue 2

Situation : Visiting the production line.

David : Before we enter the factory, please put on your helmet, Mr. Green.

Mr. Green : All right. Do I need to put on the jackets, too?

David : Yes, it's better for your clothes. Now this way, please.

Mr. Green : Is the production line fully automated?

David : No, not fully automated at present.

Mr. Green : I know. How do we control the quality?

David : During the whole manufacturing process, there are five checks all together to guarantee the quality of products.

Mr. Green : What's the monthly output?

David : Now the output is about 1000 units per month. And the number will be 1300 starting from September.

Mr. Green : What about the percentage of rejects?

David : About 4% in normal operations.

Mr. Green : Oh, that's really great.

David : Mr. Green, let's take a look at another workshop, shall we?

Mr. Green : Certainly.

Notes: This way, please：请走这边。way 意思是"路，路径"，"前往……的路"是"the way to..."。

all together：一共，总共。

Task 3　Find the correct answer for Speaker A.

Speaker A	Answer
1. Hi, Chen Gang, what are you busy with?　_____	a. Sure, here is the key.
2. May I use your bicycle this afternoon?　_____	b. From your advertisement on the Internet.
3. How was your journey to Beijing, Liu Hui?　_____	c. An apartment near my office.
4. How did you get to know our product?　_____	d. It's great.
5. What kind of apartment do you want to rent?　_____	e. I'm writing a production plan.

Task 4　Look at the pictures, and choose the correct English expression for each item.

1. 电容器 (　　)　2. 二极管 (　　)　3. 发光二极管 (　　)

4. 电阻器 (　　)　5. 磁铁 (　　)

A. diode　　B. capacitor　　C. magnet　　D. resistor　　E. LED

Part Two　Let's Read

Ⅰ. Pre-reading Activities.

Task 1　Read aloud the following words and learn the meanings.

exporter [ˈekspɔːtə(r)] *n.* 出口商	心诚意地
volume [ˈvɒljuːm] *n.* 量；体积	goods [ɡʊdz] *n.* 商品
industry [ˈɪndəstri] *n.* 产业；行业	reliable [rɪˈlaɪəb(ə)l] *a.* 可靠的；真实
enclose [ɪnˈkləʊz] *v.* （随信）附上	可信的
chamber [ˈtʃeɪmbə(r)] *n.* 议会；议院	satisfy [ˈsætɪsˈfaɪ] *v.* 令人满意
faithfully [ˈfeɪθfəli] *ad.* 忠实地；诚	catalog [ˈkætəlɒɡ] *n.* 目录，目录册
	assistance [əˈsɪstəns] *n.* 帮助，援助

Task 2　Write the words according to the English explanation
（Hints：the words from task 1. ）

1. _____ make happy or satisfied.
2. _____ put something with a letter.
3. _____ things that are made to be sold.
4. _____ in a faithful manner.
5. _____ worthy of reliance or trust.
6. _____ aid or help.

Task 3　Match the English words with the correct item.

_____ 1. well-known　　　A. 财政的

_____ 2. expand　　　　　B. 副本

_____ 3. list   C. 联系

_____ 4. experience   D. 知名的

_____ 5. copy   E. 扩大

_____ 6. credit   F. 经验

_____ 7. financial   G. 信用

_____ 8. contact   H. 清单

II . Reading.

Dear Mr. Black,

We are well-known exporters of **all kinds of**[1] Chinese goods especially mechanical products.

We expect that[2] you could provide us with a list of reliable business which are interested in our products, in order to expand the volume of business in your country. And we would be very appreciated. With our experience in the mechanical industry for more than 30 years, we are sure to make our customers satisfied.

Enclosed is a copy of our recent product catalog.

For our credit and financial situation, please contact **Bank of China**[3] and the **Chamber of Commerce in Changchun**[4].

Thank you very much for your assistance. We earnestly look forward to your reply.

Yours faithfully,

Mr. Zhou Xiang

Notes:
1. all kinds of 意思是 "各种各样的"。
2. we expect that... 表示 "我们期待……"，that 后接从句。
3. Bank of China：中国银行。
4. Chamber of Commerce in Changchun：长春商会。

阅读参考译文

III. After-reading Activities.

Task 4　Answer the questions according to the passage.

1. Who wrote this letter? What did he / she do?
2. What was the purpose of this letter?
3. What did the exporters sell?
4. Did this letter enclose a copy of our recent product catalog?
5. What's the Chinese meaning of "Chamber of Commerce"?

Task 5　Fill in the blanks according to the passage.

We are well-known 1. ＿＿＿＿＿＿ of all kinds of Chinese goods especially mechanical products.

In order to expand the 2. ＿＿＿＿＿＿ of business in your country，we expect that you could 3. ＿＿＿＿＿＿ us with a list of reliable business which are interested in our products. And we would be very 4. ＿＿＿＿＿＿. With our experience in the mechanical industry for more than 30 years，we are sure to 5. ＿＿＿＿＿＿ our customers satisfied.

Task 6　Translate the sentences into Chinese.

1. We are well-known exporters of all kinds of Chinese goods especially mechanical products.
2. We expect that you could provide us with a list of reliable business which are interested in our products.
3. We are sure to make our customers satisfied.
4. Enclosed is a copy of our recent product catalog.
5. We earnestly look forward to your reply.

Task 7   Complete the following sentences by choosing an appropriate answer from the box.

a list of   provide   more than   enclose   assistance   reply

1. This island existed _____ one hundred years ago.
2. Please notice everyone according to _____ user names.
3. The salesman _____ the customers with a new product.
4. I am looking forward to your _____.
5. I _____ two cards along with this letter.

Task 8   Translate the following sentences into English.

1. 我们是中国知名的演说家。（speaker）

   _____

2. 公司为每个人提供了工作服。（provide）

   _____

3. 我们对你们的机械产品很感兴趣。（interest）

   _____

4. 随信附上一份目录清单。（catalog）

   _____

5. 非常感谢你们的援助。（assistance）

   _____

⟳ Part Three   Let's Write

Telephone Message

电话留言属于便条的一种，在日常办公环境中应用得很多。电话留言的格式与其他便条的格式基本一致，但是有些公司有固定的电话留言条，以一种填表的形式记录电话留言。

电话留言通常包括以下几个基本要素：

（1）电话留言对象的姓名（称谓）。

（2）电话留言人的姓名（称谓）。

（3）来电日期和时间（日期可省）。

（4）留言的内容或要求。

（5）来电人联系方式（主要是电话号码或电子邮箱）。

下图为电话留言的常用格式（也有用便条格式的留言条，并没有硬性规定）：

```
                    Telephone Message

    From:  _____          To:  _____
    Date:  _____          Time:  _____
    Message:  _____
              _____
    Signature:  _____
```

Sample

```
                    Telephone Message

    From: Jessie          To: Lucas
    Date: Jan.25          Time: 10:00 a.m.
    Message: Jessie wants you to send the new sales plan to her through
             email at Jessie1002@163.com as soon as you come back.
    Signed by: Ruby
```

公司电话留言条固定格式如下：

Telephone Message	
Date：	Time：
From：	To：
TELEPHONED *	PLEASE CALL BACK *
CALLED TO SEE YOU *	WILL CALL AGAIN *
WANTS TO SEE YOU *	URGENT *
Message：	
Signed by：	

另有书信式留言条，例文如下：

> April 20th, 2018
>
> Dear Yang,
>
> Mr. Moore called you this morning and he wanted to meet you at your lunch time on Friday. If that is not convenient for you, please call him back today before 6:00 p.m. or any time tomorrow. His number is 863798.
>
> Yours,
>
> Clair

Useful Expressions

1. Please call him as soon as you come back.
   请你一回来就给他打电话。
2. Please give her a call back upon seeing the message.
   看到信息请给她回电话。

3. He wants to know if this is a good timing for you to meet him.
他想知道这个时间您是否方便见面。

4. Please call him back today before 6:00 p. m. or any time tomorrow.
请于今天下午6:00前或明天任何时候回电话给他。

5. She wanted to discuss the sales plan for next quarter.
她想和你讨论下个季度的销售计划。

6. Please give him a reply as soon as possible. 请尽快给他回复。

7. Ruby said she can't see you this afternoon.
鲁比说她今天下午不能和你见面。

8. Mr. Carter called to tell you that he will look into the matter in person.
卡特先生来电说他会亲自调查此事。

9. They hope you can send the catalog to them as soon as possible.
他们希望你能尽快把目录寄给他们。

10. Mary called that she can't go to your party.
玛丽打电话说她不能去参加你的聚会。

Writing Practice

A. *Fill in the blanks.*

说明：按电话留言的格式和要求，以秘书 Anne 的名义给 Mr. Hunter 写一份电话留言，内容如下：

1. 日期：2018 年 11 月 10 日。

2. 来电时间：上午 10 点。

3. 来电人：ABC 公司（ABC Company）的 Jennie。

4. 事由：Mr. Wilson 明天下午 5:00 从北京来上海，按原计划与 Mr. Hunter 商讨合作项目细节。

_____(1)_____

Dear _____(2)_____,

Jennie from ABC Company called you _____(3)_____（2018 年 11 月 10 日上午 10:00）. Mr. Wilson will come to Shanghai from Beijing and he will _____(4)_____（讨论合作的细节）according to the original schedule.

_____(5)_____

(1) _____

(2) _____

(3) _____

(4) _____

(5) _____

B. *Write a message according to the following requirements.*

说明：NEC 公司的营销部经理 Mr. Carter 打电话给北京分公司经理李阳。李阳外出出差，秘书张晓丽接的电话，电话内容是：总公司准备圣诞节期间在北京举行产品促销活动，要求北京分公司与北京各商场取得联系，安排好下个月的促销活动。

来电时间：2018 年 11 月 3 日上午 9：00。

Words for reference：

营销部：Marketing Department

促销：sales promotion

Part Four   Learning More

I . Vocabulary

表示各种图的相关词汇

■ **illustration** *n.* 插图，图解

illustration design 插图设计

Function of the machine can be seen in the illustration below.

这台机器的功能请见下面图解。

■ **picture** *n.* 图画，图像

The little girl is drawing a picture with coloured pencils.
小女孩正在用彩色铅笔画一幅图画。

■ **diagram** *n.* 图标，简图

block diagram 框图　circuit diagram 电路图　connection diagram 接线图
data flow diagram 数据流程图　flow diagram 流程图

■ **view** *n.* 视图

front view 主视图　lateral view 侧视图　plan view 平面图

■ **sketch** *n.* 草图，图样

There is a sketch of building site. 有一张建筑工地的草图。

■ **graph** *n.* 曲线图，图表

See the graph above and write your explanation. 看上面的曲线图并写出
你的解释。

■ **outline** *n.* 轮廓图，略图

He drew the outline of his school. 他画出他学校的轮廓图。

Practice

Ⓐ *Read the following sentences and underline the related words that mean "图".*

1. The graph showed that the increase in trade between the years 2000 and 2010.

2. There are some coloured illustrations in the book.

3. Look at a lateral view.

4. Can you draw a flow diagram?

5. We need a sketch of the device.

B. Translate the five sentences above into Chinese.

1. _____

2. _____

3. _____

4. _____

5. _____

Ⅱ. Grammar

介　词

介词又称作前置词，表示名词、代词等与句中其他词的关系，在句中不能单独作为句子成分。介词通常位于名词或代词之前。

常用的介词有 in，on，with，by，for，at，about，under，of，into，within，throughout，inside，outside，without 等，还有一些介词短语，如 because of，according to，in addition to，in front of，in spite of，by means of 等。

介词按其常用用法有如下分类：

（1）表示时间，常用介词 at，on，in，during，for，over，within，throughout，from，to 等，还有 before，after，since，until，till，between，up to 也能表示时间概念。

例如：in March（月份）；in 2002（年份）；in winter（季节）

at ten o'clock（表示某一钟点）；at first 首先

on Monday 周一；on Monday morning 周一早上

for one hour 一个小时；for a while 一会儿

（2）表示方位，常用介词 at，in，on，to，for，above，over，on，below，under，in front of，in the front of，at the back of，beside，behind，beside，across，around 等。

例如：on the table 在桌子上；above my head 在我头顶上

over the river 在河上方；below the line 在线下方

in front of the car 在车前面；in the front of the car 在车的前部

（3）表示方式手段，常用介词 by，in，with，through，by means of 等。

例如：write with a pencil 用铅笔写字；in this way 用这种方式

by bike 骑车；in English 用英语

（4）表示原因，常用介词 for，from，of，through，because of，due to，for fear of，on account of，out of 等。

例如：He came here for a book. 他来这是为了借一本书。

The flight was postponed on account of the weather.

由于天气关系，航班延误了。

The price of vegetables is going high because of the flood.

由于洪水的原因蔬菜的价格持续上涨。

（5）表示目的，常用介词 for，on，for the purpose of，for the sake of 等。

例如：He runs five kilometers every day for his health.

他为了身体健康每天跑步 5km。

My father stopped smoking for the sake of his health.

为了健康我爸爸戒烟了。

（6）表示条件，常用介词 without，but for，in case of，in the event of 等。

例如：But for your help I can't finish the work today.

如果没有你的帮助我今天不能完工。

In the event of fire, ring the alarm bell. 如果着火，就按警铃。

Practice

Fill in the blanks with proper preposition.

1. They went to the park _____ foot yesterday because they missed the bus.

2. The event was _____ the morning of Oct. 19th.

3. He bought the house _____ a good price.

4. They have lived here _____ thousands of years.

5. Linda will finish her work _____ two days.

6. There is a picture _____ the sofa _____ the living room.

7. There is a blackboard _____ the classroom.

8. The teacher came into the classroom _____ a book _____ her arm.

9. The ice will change _____ water when the temperature is _____ 0℃.

10. The manager will leave _____ New York _____ a conference.

Unit Five
Reservation

 Part One Let's Talk

Task 1 Read the sentences aloud and do exercises.

■ I'd like to make a reservation for...我想预订……
  e. g. I'd like to make a reservation for two nights.
  我想预订一间房间，住两晚。

■ Have you got any vacancies for...? ……有空房间吗？
  e. g. Have you got any vacancies for the nights of 12th and 13th?
  12 和 13 号有空房间吗？

■ We have... available. 我们有……还未被预订。
  e. g. We have a single room available.
  我们有一间单人房还未被预订。

■ How much do you charge for...? ……多少钱？
  e. g. How much do you charge for one night? 多少钱一晚？

■ Can I have your…, please? 可以告诉我您的……吗？

　　e. g. Can I have your name and contact number, please?
　　可以告诉我您的姓名和联系电话吗？

Exercise **1**　Fill in the blanks.

1. 我想预订一间单人间。
   I'd like to make a _____ for a _____ _____.

2. 本周末有空房间吗？
   Have you got any _____ for the _____?

3. 我们有一间双人间还未被预订。
   We have a _____ room _____.

4. 住一天多少钱？
   How much do you _____ for _____?

5. 可以告诉我您的姓名和地址吗？
   Can I _____ your _____ and _____, please?

Exercise **2**　Rearrange the sentences to make a dialogue.

A. Good morning, I'd like to make a reservation for this weekend.

B. A single room, please.

C. Good morning, Langyue Hotel. What can I do for you?

D. 298 RMB for one night.

E. All right. Would you like a single room or a double room?

F. Let me see. We have a single room available for 15th and 16th.

G. OK. Here's my ID card.

H. That's great. How much do you charge for one night?

_____ _____ _____ _____ _____ _____ _____ _____

Task 2   Role-play.

> ### Words bank
>
> vacancy 空缺   make a reservation 预订   single room 单人间
> contact number 联系电话   equipment 设备   catalogue 目录
> price 价钱   order 订购

Dialogue 1

Hotel Clerk：Good morning, sir, Hilton Hotel. May I help you?

Mr. Gao　: Good morning. Have you got any vacancies tomorrow night? I'd like to make a reservation for one night.

Hotel Clerk：All right. Single room or double room?

Mr. Gao　: Single room, please.

Hotel Clerk：Wait a moment, let me check. Yes, we have a single room available.

Mr. Gao　: How much do you charge for one night?

Hotel Clerk：A single room is 80 dollars per night.

Mr. Gao　: Do you accept VISA?

Hotel Clerk：Yes, we do. Can I have your name and contact number, please?

Mr. Gao　: Yes, it's Gao Dong. My number is 446-1227.

Hotel Clerk：All right, Mr. Gao. A single room for tomorrow night. We look forward to your visit then. Have a nice day!

Mr. Gao　: Thank you.

Dialogue 2

Susan : Good afternoon, Miss Chen. I see that you have read our catalogue. Is there any equipment you are interested in?

Miss Chen: Yes, we are quite interested in your boring-milling machine. But we found that the price is much higher than we expected.

Susan : How many would you like to order?

Miss Chen: 20, if the price is appropriate. But with the price on the list now, we won't make profit. Can we negotiate on that?

Susan : Um, all right.

Miss Chen: Thank you. When can you make the goods ready for shipment?

Susan : July.

Miss Chen: Well, we expect them to be sent by September. The goods need to be transmitted in the middle of the shipment. So could you get the goods ready before June?

Susan : Yes, no problem.

Miss Chen: Good. When can we sign the contract?

Susan : This afternoon.

Miss Chen: OK. See you later.

Susan : See you.

Task 3   Find the correct answer for Speaker A.

Speaker A	Answer
1. Why didn't you buy the mobile phone you like? _____	a. At 16:30.
2. When is the train leaving for Beijing? _____	b. I've got a headache with a slight fever.
3. When are you going to take the new job? _____	c. She has lost her job.
4. You don't look well. What's wrong with you? _____	d. As soon as possible.
5. Mary looks worried. What has happened to her? _____	e. It's too expensive.

Task 4   Look at the picture, and choose the correct English expression for each item.

1. 陀螺仪（   ）   A. antenna
2. 速度传感器（   ）   B. camera
3. 机器人关节（   ）   C. gyroscope
4. 天线（   ）   D. speed sensor
5. 机器人手臂（   ）   E. joint
6. 摄像机（   ）   F. battery
7. 机身（   ）   G. arm
8. 电池（   ）   H. body

Part Two　Let's Read

Ⅰ. Pre-reading Activities.

Task 1　Read aloud the following words and learn the meanings.

reprogrammable [riˈprəʊɡræməbl] *a.* 可改编程序的	manipulate [məˈnɪpjʊˌleɪt] *v.* 操作，处理
motion [ˈməʊʃ(ə)n] *n.* 运动；动机	derive [diˈraɪv] *v.* 从……衍生；起源于
invent [inˈvent] *v.* 发明	intelligent [ɪnˈtelɪdʒ(ə)nt] *a.* 智能的
possess [pəˈzes] *v.* 拥有	sensory [ˈsensəri] *a.* 感官的；传递感觉的
perception [pəˈsepʃ(ə)n] *n.* 知觉；觉察	artificial [ˌɑːtɪˈfɪʃ(ə)l] *a.* 人造的；人工的
ascribe [əˈskraɪb] *v.* 把……归于；认为……具有	

Task 2　Write the words according to the English explanation
（Hints：the words from task 1.）

1. _____ to own sth.
2. _____ to control or influence.
3. _____ to make or design sth. that doesn't exist.
4. _____ produced or made by mankind.
5. _____ the act or movement.
6. _____ able to learn or understand or think sth. with wisdom.

Task 3　Match the English words with the correct item.

_____ 1. intelligent robot　　　　　A. 机器人应用

_____ 2. automatic device   B. 智能机器人

_____ 3. robotics   C. 视觉传感器

_____ 4. task   D. 触觉传感器

_____ 5. visual sensor   E. 人工智能

_____ 6. tactile sensor   F. 任务

_____ 7. robot application   G. 机器人学

_____ 8. artificial intelligence   H. 自动化设备

Ⅱ. Reading.

Industrial Robots

"A robot is a reprogrammable, multifunctional machine designed to manipulate materials, parts, tools, or specialized devices, through variable programmed motions for the performance of a variety of tasks."

——Robotics Industries Association

"A robot is an automatic device that performs functions normally ascribed to humans or a machine in the form of a human."

——Webster's Dictionary

Introduction

The word "robot" was derived from the Czech word which means forced labor. The study of robots is referred to as "**robotics**[1]". **In late 1950s**[2] and early 1960s, Geroge Devil and Joe Eagleburger invented the first modern industrial robots, which were named the Animates. Then Eagleburger established an organization called the Animation. Since he was the first to make robots, he was called the "father of robots".

Intelligent Robots

The emerge of "**intelligent robots**[3]" has brought the development of robot application **to a new stage**[4]. Intelligent robots refer to the ones that have intelligence and are capable of reacting to their surroundings. Thus they must possess sensory perception and artificial intelligence. Research in robotics has always been around the equipment of eyes and fingers of robots, also known as visual sensors and tactile sensors. The applications of artificial intelligence make the robot capable of responding and adapting to its task, **as well as**[5] its environment. And also owing to artificial intelligence, the intelligent robots are smart enough to reason and make decisions according to the changes in its surroundings.

Notes:

1. robotics：robotic 的复数，也指机器人学。
2. in late 1950s：在二十世纪五十年代晚期，如果只表示年代，则有定冠词 the，如 in the 20s.
3. intelligent robot：智能机器人，industrial robot 是工业机器人。
4. bring... to a new stage：将……带入到新阶段。
5. as well as：也，同……一样。

阅读参考译文

Ⅲ. After-reading Activities.

Task 4　Answer the questions according to the passage.

1. What does "robotics" refer to?
2. When were the first modern industrial robots invented?
3. What are intelligent robots?
4. What's the main study in robotics?
5. What are the advantages of the application of artificial intelligence?

Task 5　Fill in the blanks according to the passage.

The emerge of "intelligent robots" has brought the development of robot application to a new stage. 1. _____ robots·refer to the ones that have intelligence and are capable of reacting to their 2. _____. Thus they must possess sensory perception and 3. _____ intelligence. Research in robotics has always been around the equipment of eyes and fingers of robots, also known as 4. _____ sensors and 5. _____ sensors. The applications of artificial intelligence make the robot capable of responding and adapting to its task, as well as its environment.

Task 6   Translate the sentences into Chinese.

1. The study of robots is referred to as "robotics".
2. Since he was the first to make robots, he was called the "father of robots".
3. The emerge of "intelligent robots" has brought the development of robot application to a new stage.
4. Intelligent robots refer to the ones that have intelligence and are capable of reacting to their surroundings.
5. The application of artificial intelligence make the robot capable of responding and adapting to its task, as well as its environment.

Task 7   Complete the following sentences by choosing an appropriate answer from the box.

| surroundings   invent   intelligent   derive   application   adapt to |

1. Mike is a very _____ child. He has been exceptionally admitted into college this summer.
2. After lights were _____, people were able to work and study into the nights.
3. Students can do their self-study of lessons through the new _____ software on their phones.
4. Everyone likes to work in pleasant _____.
5. Graduates always find it hard to _____ the society.

Task 8　Translate the following sentences into English.

1. 中秋节也常被称作月饼节。（refer to…as）

   _____

2. 这项技术仍处于开发阶段。（stage）

   _____

3. 当地政府对这则坏消息表示非常愤怒。（react to）

   _____

4. 智能机器人能够感知周围的环境。（capable of）

   _____

5. 由于资金缺乏，该项目将被取消。（owing to）

   _____

◯ Part Three　Let's Write

Invitation

Invitation 可以分为邀请信（Invitation Letter）和请柬（Invitation Card）。

邀请信一般适合用于某些普通场合，多用于亲朋好友之间的邀请。请柬广泛用于公司开业典礼、婚礼、宴会、晚会等正式社交场合。写请柬的时候，应该写明受邀请者的姓名、受邀请的理由、日期和地点等。二者的主要区别在于：

（1）请柬应该用第三人称写，而邀请信用第一人称写。

（2）请柬内容一般居中书写，邀请信的格式和书信格式相同。

邀请信的格式和一般英语书信一样，可以用缩进式、齐头式和混合式。但是其内容主要分为三部分：

第一段写明目的。

第二段提供主要信息，比如时间、地点等。

第三段再次强调邀请愿望。

同时，开头和结尾都要有相应的呼语和署名、日期。

而请柬没有统一固定的格式要求，除了要把信息表达清楚外，对其美观性有较高的要求。

Sample 1　Invitation Letter

Sept.10, 2018

Dear Sir or Madam,

　　We would like to linvite you to attend the 2018 International Trade Fair, which will be held from October 8 to 20 in Guangdong. Further details on the fair will be sent to you later.

　　We look forward to hearing from you soon, and we would appreciate having your acceptance.

Yours faithfully,

Richard Baldwin

收到邀请信后，不管是接受还是拒绝，都要给对方写信回复，这在英语国家尤其重要。如果接受邀请，写信的内容主要是：

（1）感谢对方的邀请。

（2）愉快地表示接受。

（3）表示期待赴会和与对方见面的心情。

如果因为某些原因不能接受邀请，也就是拒绝邀请，写信的内容主要是：

（1）依然表示感谢。

（2）说明原因，对无法出席表示遗憾。

（3）表达祝愿，即预祝活动顺利进行。

邀请信的回信中应明确表明接受邀请还是不接受邀请，不能含糊其辞，例如不能写"I'll come if I'm in town"这类的话，以使对方无法做出安排。在接受邀请的复信中，应对受到邀请表示高兴。谢绝的回信中应阐明不能应邀的原由。

Sample 2   Accepting Letter

Sept. 13, 2018

Dear Mr.Baldwin,

Thank you for your letter of September 10 inviting us to participate in the 2018 Intemational Trade Fair. We are very pleased to accept the invitation and will plan to display our smart home appliances.

Mr. Cook will be in your city from October 2 to 8 to make arrangements and would very much appreciate your assistance.

Yours faithfully,
Mary Anderson

Sample 3   Declining Letter

Sept. 11, 2018

Dear Mr. Baldwin,

Thank you very much for your invitation to attend the 2018 International Trade Fair. As we are going to open an aftersales shop in your city at that time, we are sorry that we shall not be able to come.

We hope to see you on some future occasion.

Yours faithfully,
Sarah Miller

Useful Expressions

1. This letter is to invite you to attend the conference of...
    这封信是邀请您参加……会议。

2. I would like to invite you to... 我想邀请您……

3. Please confirm your schedule as early as possible. 请您早日定下行程。

4. We sincerely hope you can join us for the event.
    我们真诚希望你能参加我们的活动。

5. Thank you very much for inviting me to your party.
    非常感谢您邀请我参加你的聚会。

6. I am pleased to come. 我很愿意参加。

7. I'd love to come but I have another appointment.
    我很愿意参加，但我另有约会（不能接受您的盛情邀请）。

8. We have pleasure in inviting you to our annual conference.
    很高兴邀请您参加我们的年会。

9. I will call you in a week or so to follow up on this.
    我会在一周左右给您打电话跟进此事。

10. It was very kind of you to ask me, but I am afraid that I will not be able
    to come.
    非常感谢您能邀请我，但是恐怕我不能去。

Writing Practice

A. *Fill in the blanks.*

说明：假设你是周维，你给你的朋友 Anne 写一封邀请信，邀请他来参加自己 6 月 8 日晚上 6:00 的生日聚会。

写信时间：6 月 2 日

　　　　(1)　　　　

Dear Anne,

　　　　(2)　　　　（我想知道你是否能来）my 23rd birthday party at my

house on this Saturday night, June 8th. It would be my pleasure to share the important moments with you. In addition, since you are a fan of rock music, _____(3)_____（我很高兴地告诉你）that I have invited our campus rock band, the "Rockbreak", to perform. There are also arrangements for singing, dancing and cake-cutting, which I _____(4)_____（相信你一定会玩得很开心）.

The dinner starts at 6:00 p.m. so that we can have a nice and long evening _____(5)_____.

_____(6)_____（我非常希望您能够光临我的晚会，请告知我你的决定。）

Love,
Zhou Wei

(1) _____

(2) _____

(3) _____

(4) _____

(5) _____

(6) _____

B. *Write an invitation letter according to the given situation.*

说明：周阳在面试之后成为市场部经理 Mr. Smith 的助理。一天，Mr. Smith 让她写邀请信邀请销售经理 Mr. Ford 在 11 月 18 日到乡间俱乐部共进晚餐。

Words for reference：

市场部经理：Marketing Manager

乡间俱乐部：Country Club

Part Four   Learning More

Ⅰ. Vocabulary

表示"最大值""最小值"的相关词语

■ **maximum** *n.* 最大值，最大数   *a.* 最多的，最高的

maximum power 最大功率   maximum speed 最高速度
maximum size 最大尺寸   maximum performance 最大性能
maximum length 最大长度

■ **maximize** *v.* 把……增加扩大到最大限度；使……达到最高限度
Please maximize a window on a computer screen.
请把计算机屏幕上的窗口最大化。

■ **minimum** *n.* 最小量，最小数，最小值   *a.* 最小的，最低的
I will take a minimum of one hour for doing sports.
我最少要用1h做运动。

■ **minimize** *v.* 使……减到最小；使……达到最低限度
We should minimize the risk of shipping.
我们应该使海运的风险减到最小。

■ **the upper limit** 上限，最高限度，最大尺寸
exceed the upper limit 超过上限

■ **the lower limit** 下限，最低限度，最小尺寸
the lower limit value 下限值

Practice

Ⓐ *Read the following sentences and underline the related words that mean* "最大值" *or* "最小值".

1. For hearing, 20Hz is the lower limit of frequency and 20,000Hz is the

upper limit.

2. The loss can be reduced to a minimum by the use of a new material.

3. Don't minimize a window on a computer screen.

4. What is the maximum size of a sofa?

5. We can use the resources to minimize the damage.

B. *Translate the five sentences above into Chinese.*

1. _____

2. _____

3. _____

4. _____

5. _____

II . Grammar

数词的写法和读法

1. 时间的写法和读法

（1）可以直接按照表示时间的数字来读。

（2）正点后的前半小时，通常说几点"过"（past）几分。

（3）正点后的后半小时，通常说几点"差"（to）几分。

（4）英语中的 15min 也可以说成"一刻钟"（a quarter）

例如：08：00 eight o'clock 或者 eight

09：15 nine fifteen 或者 a quarter past nine（美语：after nine）

02：30 two thirty 或者 half past two（美语：after two）

05：45 five forty-five 或者 a quarter to six（美语：of six）

07：00 seven hundred hours = seven a. m.

12：00 twelve hundred hours = midday = noon

14：15 fourteen fifteen = two fifteen p. m.

23：05 twenty-three oh five = eleven five p. m.

24：00 twenty-four hundred hours = midnight

2. 日期的写法和读法

Format	British：day-month-year	American：month-day-year
A	the twenty-ninth of March, 2017	March the twenty-ninth, 2017
B	29th March 2017	March 29th, 2017
C	29 March 2017	March 29, 2017
D	29/3/2017	3/29/2017
E	29/3/17	3/29/17

读法举例：比如2018年3月22日，美语可写作"March 22, 2018"。

读成 March（the）twenty-second, twenty eighteen（美语偏爱这种读法，且常省略 the），或读成 the twenty-second of March, two thousand and eighteen。

英式可写作"22 March 2018"，读成 the twenty-second of March two thousand and eighteen，或读成 the twenty-second of March, twenty eighteen。

3. 分数和小数的读法

1/2 可以读为 one half 或 one over two，1/3 是 one third，1/4 是 one fourth。在日常生活中，用 one quarter 表示 1/4 比用 one fourth 更多一些。2/3要说 two thirds，也就是说分母要加 s。

6.27 可以读作 six point twenty-seven 或 six twenty-seven。在美国买东西都要含税，所以价钱多半都带有小数点，通常小数点可以说 point，也可以直接省略。另外比较正式的说法为 six dollars and twenty-seven cents，但是在一般日常生活中几乎是听不到这种读法，而是直接读为 six twenty-seven。

Practice

Read the following time, date and other numbers in English.

12:45	4:05	13:28	11:15
4/5	2/3	3/4	5/8
1.23	14.31	6.19	7.54

2018 年 4 月 3 日   1978 年 9 月 20 日

Unit Six
Appointment

Part One   Let's Talk

Task 1   Read the sentences aloud and do exercises.

- Would you like to...? 您愿意……吗?

  e. g. Would you like to have lunch together at Garden Hotel next Wednesday?
  您下星期三可以和我在花园酒店一起吃午餐吗?

- Does... suit you? 您……有空吗?

  e. g. Does Friday suit you? 您星期五有空吗?

- Is... convenient for you? 您……方便吗?

  e. g. Is 12 o'clock convenient for you? 您12点方便吗?

- I'm afraid I can't make...我……恐怕没时间。

  e. g. I'm afraid I can't make Wednesday.
  我星期三恐怕没有时间。

- How about...? ……怎么样?

  e. g. How about Friday afternoon? 星期五下午怎么样?

Exercise 1   Fill in the blanks.

1. 您星期六晚上可以和我一起吃晚餐吗？

   Would you like to _____ together on _____ _____?

2. 下星期一您有空吗？

   Does _____ _____ suit you?

3. 星期五您方便吗？

   Is Friday _____ for you?

4. 我星期五恐怕没有时间。

   I'm _____ I can't _____ Friday.

5. 这周末怎么样？

   How about _____ _____?

Exercise 2   Rearrange the sentences to make a dialogue.

A. Oh，that's great. What time?

B. Hi，Lily. Would you like to have dinner together this weekend?

C. Is Saturday convenient for you?

D. Sure. See you then.

E. I'm sorry. I can't make Saturday. How about Sunday?

F. Sunday is OK. Does 5：30 suit you?

_____ _____ _____ _____ _____ _____

Task 2   Role-play.

Words bank		
order 订单	schedule 日程安排	conference 会议
convenient 方便的	look forward to 期待	major 主要的
responsibility 责任	eyesight 视力	position 岗位
opportunity 机会		

Dialogue 1

Betty: Good afternoon.

Jones: Good afternoon, this is Jones. May I speak to Betty?

Betty: Speaking, please.

Jones: I'd like to discuss the new order with you. Would you like to have lunch together at Garden Hotel next Wednesday?

Betty: Let me check my schedule. Er…I'm afraid not. I've got to go to Beijing for a conference. I'll come back next month.

Jones: That's a pity. Does Friday suit you?

Betty: Yeah, that would be fine. What time?

Jones: Is 12 o'clock convenient for you?

Betty: Yes. This Friday at 12:00 at Garden Hotel. See you then.

Jones: See you.

> Notes: Yeah: 口语表达法，意思同 Yes。
> Er: 呃，表示犹豫。
> I'm afraid not: 恐怕不行。作为客气的口语表达，用于委婉地否定，纠正和驳斥对方。
> what time: 相当于 when。

Dialogue 2

Dina: Mr. Black, would you like to come to Chinese Trading Fair of Machine Tools next month?

Black: Thank you, Dina. I will be there. It seems to be a big show.

Dina: Yes. There are already over four thousand people registered for this event, and the number is expected to increase in the coming days.

Black: Well, I'm really lucky to have this opportunity.

Dina : The event has been very important to machine tools industry here in China. There are about 1,200 new machine tools on display.

Black: Fantastic! Do you hold big event like this every year?

Dina : Yes, we have been holding this in China every two years since in 2010, and it has been recognized as one of the top three marketing activities in the world's machine tool area.

Black: I'm sure to come next month.

Dina : You are welcome!

> **Notes**: card：也就是 name card，名片。
> trading fair：贸易会。
> one of the top three：最好的三个中之一。

Task 3　Find the correct answer for Speaker A.

Speaker A	Answer
1. How much does the T-shirt cost? _____	a. It's over there.
2. Where can I find an application form, sir? _____	b. 20 dollars.
3. Could I have your business card? _____	c. Sorry, he is in the meeting.
4. May I talk to Mr. Zhang about the work plan? _____	d. Yes, we need.
5. Does your company still need a secretary? _____	e. Here you are.

Task 4 Look at the pictures, and choose the correct English expression for each item.

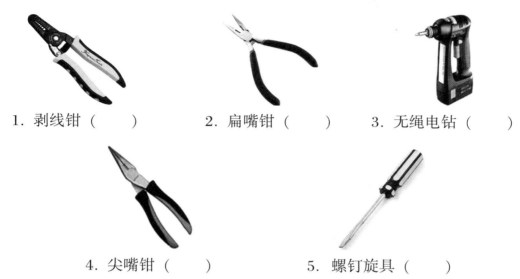

1. 剥线钳 (   )   2. 扁嘴钳 (   )   3. 无绳电钻 (   )

4. 尖嘴钳 (   )   5. 螺钉旋具 (   )

A. cordless drill B. snipenose pliers C. screw driver D. wire strippers E. flat nose pliers

Part Two   Let's Read

Ⅰ. Pre-reading Activities.

Task 1 Read aloud the following words and learn the meanings.

essentially [ɪˈsenʃ(ə)li] *ad.* 本质上地，根本上地

industrial [ɪnˈdʌstrɪəl] *a.* 工业的，产业的

keyboard [ˈkiːbɔːd] *n.* 键盘

activate [ˈæktɪˌveɪt] *v.* 使活动，使开始作用

ladder [ˈlædə(r)] *n.* 梯状物；梯子

output [ˈaʊtˌpʊt] *n.* 输出

mission [ˈmɪʃ(ə)n] *n.* 使命，任务

monitor [ˈmɒnɪtə(r)] *n.* 显示器；监视器；监控器

normally [ˈnɔːm(ə)li] *ad.* 通常地，一般地

edit [ˈedɪt] *v.* 编辑

microprocessor [ˌmaɪkrəʊˈprəʊsesə(r)] *n.* 微处理器

valve [vælv] *n.* 阀；真空管

Task 2　Write the words according to the English explanation （Hints: the words from task 1.）.

1. _____ a device for controlling the flow of a liquid or gas, letting it move in one direction only.
2. _____ a screen which is used to display certain kinds of information.
3. _____ the main microchip, which controls most important functions of computer.
4. _____ an important task that people are given to do.
5. _____ it relates to the physical senses.
6. _____ a piece of equipment used for climbing up something.

Task 3　Match the English words with the correct item.

_____ 1. ladder   A. 工业的
_____ 2. edit   B. 键盘
_____ 3. valve   C. 通常地，一般地
_____ 4. output   D. 使活动
_____ 5. keyboard E. 编辑
_____ 6. normally F. 梯状物；梯子
_____ 7. activate G. 阀；真空管
_____ 8. industrial H. 输出

Ⅱ. Reading.

PLC

The PLC is short for **Programmable**[1] Logic Controller which is essentially a computer with a single mission. Most commonly used in industrial applications, it usually works without a monitor,

keyboard, and a mouse, as it is normally programmed to operate a machine or system. Factory **assembly line**[2] machinery is activated and monitored by a single PLC. A personal computer is connected to it if the PLC program needed to be edited or created. Most PLCs are programmed in a special language called **Ladder Logic**[3].

Programmable logic controllers, sometimes referred to programmable controllers, are microprocessor based units that, in essence, monitor external sensory activity from additional devices. They take in the data which reports on a wide variety of activities, such as machine performance, energy output, and process impediment. Attached motor starters, **pilot lights**[4], valves and many other devices are also controlled by PLC.

Notes: 1. Programmable 是形容词，意为可设计的，可编程的。
该段出现的 program，意为给……编写程序。另有
programmer 意思为程序设计者、程序设计器。
2. assembly line 意思是流水线，（工厂产品的）装配线。
3. Ladder Logic：梯形逻辑，是一种编程语言。
4. pilot lights 意为领航信号灯、指示灯。

阅读参考译文

‖. After-reading Activities.

Task 4   Answer the questions according to the passage.

1. What is PLC short for ?
2. Does PLC work with a monitor or keyboard?
3. How is the PLC program edited or created?
4. What is the language in programming PLCs?
5. What other devices are controlled by PLC?

Task 5   Fill in the blanks according to the passage.

Factory assembly line machinery is 1. _____ by a single PLC. 2. _____ is connected to it if the PLC program needed to be edited or created. Most PLCs are programmed in 3. _____ called Ladder Logic.

They take in the data which reports on 4. _____. Attached motor starters, pilot lights, 5. _____ are also controlled by PLC.

Task 6 Translate the sentences into Chinese.

1. The PLC is short for Programmable Logic Controller which is essentially a computer with a single mission.
2. Factory assembly line machinery is activated and monitored by a single PLC.
3. They take in the data, which reports on a wide variety of activities, such as machine performance, energy output, and process impediment.
4. Most PLCs are programmed in a special language called Ladder Logic.
5. Attached motor starters, pilot lights, valves and many other devices are also controlled by PLC.

Task 7 Complete the following sentences by choosing an appropriate answer from the box.

monitor	edit	valve	mission	output	sensory

1. She was sent on a special _____ to Africa.
2. _____ of grain increases year after year
3. The heart _____ shows he is in bad condition.
4. He studies how the brain processes _____ information, like smell, for living.
5. I used to _____ the college paper in the old days.

Task 8 Translate the following sentences into English.

1. 在这个工业区兴起了一座新城市。(industrial)

_____

2. 我们要激励青年去学习。（activate）

_____

3. 键盘和鼠标控制你的计算机。（keyboard）

_____

4. 基本上，我们只有两种选择。（essentially）

_____

5. 阀门失去控制，我们要马上关闭机器。（valve）

_____

Part Three   Let's Write

Thank-you Letter

日常生活中经常会得到别人帮助或收到别人的礼物，这种情况下礼貌的做法是通过写一封感谢信来表达谢意。同时，在收到别人的邀请、受到别人的赞扬时，都可以写感谢信。

感谢信的写法比较简单，通常可以分为三步：

（1）表达感谢并说明原因。

（2）详细叙述事情原委，强调细节。

（3）再次感谢并表达回报愿望。

感谢信的格式同普通信的格式一样，有日期、称呼、正文、结束语和署名。另外，写信人需要注意的是，感谢信一定要写得及时，不要事情已经过去一段时间了再写，这样不管内容写得如何都显得没有诚意。

Sample 1

Oct. 12, 2018

Dear Mrs. Mililer,

It was so generous of you to give my husband and me so much of your time. We enjoy and appreciate all your gracious hospitality to us more than we can see.

We hope that you and Mr. Green may come to China before long and we may have the pleasure of seeing you then in our home.

Thanks again for your kindness and my husband joins me in sending our best wishes to you and your family.

Very sincerely yours,

Kim Burton

Sample 2

Nov. 17, 2018

Dear Linda,

It was very grand of you to see me off at the airport. I know it's not easy for you to go to the airport. I'm very happy when I see you there.

Though the flight was nearly six hours, I spent enjoyable time because the book you had given to me.

I'm so lucky to have you as my friend. I'd like to express my appreciation again to your kindness to me.

Yours,

Kelly

Useful Expressions

1. Thank you very much for... 非常感谢……

2. Many thanks for your... 非常感谢您的……

3. I'm most thankful for your... 非常感谢您……

4. I am truly grateful to you for... 真心感激您……

5. I sincerely appreciate... 我衷心地感谢……

6. Please accept my sincere appreciation for...
   请接受我对……真挚的感谢。

7. You were so kind to give me your help. 承蒙您的帮助。

8. Once again millions of thanks for your... 再次对您的……表示万分感谢。

9. Thank you again for your wonderful hospitality and I am looking forward to seeing you soon.
   再次感谢您的盛情款待，并期待不久见到您。

10. Thank you for your kindness to have done me a favor.
    感谢您的帮助。

Writing Practice

A. *Fill in the blanks in the thank-you letter below with the words and expressions given.*

gratitude	appreciation	assist	express
look forward to	add	greet	helpful

Dear Ms. Costa,

Thank you very much for _____ my daughter while she was in London.

I know that she has already written to you _____ her _____, but I would like to my own _____. The preparations you made for her and the information she gained will be extremely _____.

We are _____ the pleasure of _____ you in Beijing in the near future.

Yours sincerely,

Kelly Jones

B. *Write a thank-you letter according to given situation.*

说明：周华上周因为感冒没有上课，她的同学刘萱琳帮助她补习了上周落下的功课，使得她能够顺利通过期中测验。周华现在想写一封感谢信来感谢刘萱琳的帮助。

时间：12 月 6 日

⟳ Part Four   Learning More

Ⅰ. Vocabulary

表示"距离"的相关词语

■ **distance** *n.* 距离；远处；间隔

center distance 中心距   minimum distance 最小距离

walking distance 步行距离   safe distance 安全距离

The distance from home to school is about 1000 metres.

从家到学校的距离大约是 1000 m。

■ **space** *n.* 距离，空间，间隔

space model 空间模型   public space 公共空间

free space 自由空间   disk space 磁盘空间

In the space of 10 km we can't see anything.

在 10 km 的距离内我们什么都看不见。

■ **way** *n.* 距离，路程

There is still a long way from here.

离这里还有很长距离。

■ **be away**（from...）离……远

The nearest hotel is about 2 km away（from here）.

最近的酒店离这里大约 2 km。

■ **apart** *ad.* 相距，分离着

My parents and I lived just one building apart from each other.

父母和我住得彼此仅隔一栋楼。

Practice

(A.) *Read the following sentences and underline the related words that mean* "距离".

1. In this row of trees the distance between two trees is 2 meters.

2. There is a space of 2000 meters from the start to the finish.

3. The maximum distance to the point is 55 km.

4. Is your company far away from here?

5. We lived apart from each other.

(B.) *Translate the five sentences above into Chinese.*

1. _____

2. _____

3. _____

4. _____

5. _____

Ⅱ. Grammar

分　词

分词是动词的变化形式，分为两种：现在分词（The Present Participle）和过去分词（The Past Participle）。分词在句子中可以做谓语，还可以充当其他成分，如定语、状语、表语、补足语。

现在分词的构成：doing，否定式为 not doing。现在分词具有主动意义，表示主动发出的动作，也可表示正在进行的动作。

过去分词的构成：done，否定式为 not done。过去分词具有被动意义，表示被动承受的动作，也可表示已经完成的动作。

（1）作定语

e. g. The sleeping baby is my little brother. 熟睡的宝宝是我的弟弟。

　　　　After the competition, my teacher gave me a satisfied smile.

　　　　比赛结束后，我的老师满意地笑了。

（2）做状语

e. g. The little girl came running into the house. 小女孩跑进屋子里。

　　　　Seen from the hill, the city looks really modern and beautiful.

　　　　从山上看这座城市，美丽而又现代化。

分词也可以做句子的状语：

e. g. Generally speaking, men have more promotion opportunities than women in this business.

　　　　总地来说，在这个行业中男性比女性有更多升职的机会。

　　　　Given good weather, our ship will arrive in England ahead of schedule.

　　　　如果天气好，我们的船将会提前到达英国。

（3）做表语

e. g. The novel was rather boring. 这本小说相当枯燥。

I was quite interested in the movie. 我对这部电影非常感兴趣。

（4）补足语

e. g. I saw a dog standing in front of my house.

我看见一只狗站在我家门口。

I wish all your problems settled. 我希望你所有的问题都可以解决。

注意：分词前有时会有一个逻辑上的主语，成为独立结构（Absolute Construction）。动词与逻辑主语之间的关系决定了分词的形式。

e. g. The day being fine, we decided to go for a picnic.

天气晴朗，我们决定去野餐。

Everyone, Jack excepted, agreed with the plan.

除了杰克以外，所有人都赞成这个计划。

Practice

Fill in the blanks.

1. The _____ (reduce) price will save you 50 cents for each pair.

2. Anyone _____ (travel) to another country is in need of a passport.

3. She was _____ (annoy) at your attitude.

4. _____ (invite) more than ten guests, we have to prepare sufficient food and drink.

5. The boy was lying in bed _____ (cry).

6. When _____ (find) the time was so late, I decided to stop working and go to bed.

7. All day my father worked with the door _____ (lock).

8. The professor said that he would come if _____ (invite).

9. I had my bicycle _____ (repair) last week.

10. The woman _____ (dress) in yellow is my aunt.

Unit Seven
Suggestions

Task 1   Read the sentences aloud and do exercises.

■ You should… for… 你要…… 因为……

e. g. You should thank them for coming all the way here.
你要感谢对方远道而来。

■ Could you tell me…? 你能告诉我……?

e. g. Could you tell me what to do to be suitable and polite?
你能告诉我怎么说才不会失礼吗?

■ Why don't you…? 你为何不……呢?

e. g. Why don't you begin with expressing your thanks to them first?
你为何不以表达你对对方的感谢作为开场呢?

■ We would like to… 我们应……

e. g. We would like to extend a warm welcome to you all.
我们应对各位表示热烈的欢迎。

■ Should I tell... about...? 我是否可以告诉…… 有关……?

e. g. Should I tell them about their schedule?

我是否可以告诉他们有关的行程？

Exercise 1   Fill in the blanks.

1. 你要感谢对方的来访。

   You _____ _____ them for _____.

2. 你能告诉我怎么表达感谢吗？

   _____ you _____ me _____ to express my gratitude?

3. 你为何不先给对方公司打个电话呢？

   Why _____ you _____ their company first?

4. 感谢各位的热烈欢迎。

   We _____ _____ to thank you for your warm _____.

5. 我可以给他们讲讲项目的事吗？

   Should I _____ them _____ the program?

Exercise 2   Rearrange the sentences to make a dialogue.

A. No problem. How can I help you?

B. Can you tell me how to leave somebody a good first impression?

C. Jim. I have to receive a business delegation next week. I may need your help.

D. That's a good idea. Thank you.

E. Why don't you send the schedule to them through emails? So they can check it whenever they want to.

F. You should greet them politely, then take them to their hotel and check in for them.

G. How can I let them know the schedule of the trip?

_____ _____ _____ _____ _____ _____ _____

Task 2   Role-play.

Words bank

receive 接收   business group 商务团体   suitable 合适的
express 表达   extend 扩大   injury 受伤
loose-fitting 宽松的

Dialogue 1

Sue ：Mike，may I ask you a question？

Mike：Sure．What's the matter？

Sue ：I will receive a foreign business group on Sunday．Could you tell me what to do to be suitable and polite？

Mike：All right．Firstly you can express your thanks to them．You may say "We would like to thank you for visiting our company"．

Sue ：I see．

Mike：Then you can say "I would like to extend a warm welcome to you all"．

Sue ：Should I tell them about their schedule？

Mike：Of course．You can also tell them "We will do our best to give you a comfortable visit．Please feel free to ask us any questions if you have．"

Sue ：Thank you very much！

Notes： What's the matter? 怎么了？
I see: 我明白了，see 在此不是 "看" 的意思。
a warm welcome: 表示热烈欢迎。

Dialogue 2

Carl：I know that we need things to protect ourselves when operating a machine. Can you tell me more specific?

Alan：Sure. You must wear protective equipment, such as safety glasses, goggles, gloves, hair nets, masks where required. Don't wear jewelry or loose-fitting clothing when operating machinery, because they may get caught in the mechanism and cause injury.

Carl：It's quite different from our daily living situation. I must be careful all the time while working.

Alan：Yes. Well, while working, you must never distract the attention of another person, since you might cause him or her to get injured. If you really have to get another person's attention, please wait until it can be done safely.

Carl：I'll keep that in mind.

> **Notes**：protect：保护。protect...from...意思是"防止……受到（伤害）"。
>
> get injured：受到伤害，同于其被动语态形式 be injured。

Task 3　Find the correct answer for Speaker A.

Speaker A	Answer
1. Have you bought a car? _____	a. Fine, thanks.
2. How is everything going, Jim? _____	b. I'm Zhang Ming.
3. May I have your name, please? _____	c. Yes, once a month.
4. Do you often travel on business? _____	d. All right.
5. Would you sign your name here, sir? _____	e. Yes, it's a red car.

Task 4   Look at the picture, and choose the correct English
expression for each item.

1. 扩展模块（   ）   A. proximity switch
2. 电动机启动器（   ）   B. expansion module
3. 接近开关（   ）   C. limit switch
4. 限位开关（   ）   D. push button
5. 按下按钮（   ）   E. motor starter

 Part Two   Let's Read

Ⅰ. Pre-reading Activities.

Task 1   Read aloud the following words and learn the meanings.

automation [ˌɔːtəˈmeɪʃ(ə)n] *n.* 自动化	primary [ˈpraɪməri] *a.* 首要的，主要的
standardize [ˈstændədaɪz] *v.* 使标准化	technique [tekˈniːk] *n.* 技术；技艺
replace [rɪˈpleɪs] *v.* 取代，代替	inconsistent [ɪnkənˈsɪst(ə)nt] *a.* 不一
quality [ˈkwɒliti] *n.* 质量	致的
device [dɪˈvaɪs] *n.* 装置，设备	process [ˈprəʊses] *n.* 过程
accomplish [əˈkʌmplɪʃ] *v.* 完成；达到	feedback [ˈfiːdbæk] *n.* 反馈；反应
（目的）	sensing [ˈsensɪŋ] *n.* 感觉；指向

Task 2   Write the words according to the English explanation
（Hints: the words from task 1. ）

1. _____ not in agreement.
2. _____ put in effect.

3. _____ take the place or move into the position of.

4. _____ the act of implementing the control of equipment with advanced technology.

5. _____ of first rank or importance.

6. _____ response to an inquiry or experiment.

Task 3 Match the English words with the correct item.

_____ 1. control system   A. 劳动力
_____ 2. upgrade   B. 因素
_____ 3. labor   C. 控制系统
_____ 4. energy   D. 提升
_____ 5. accuracy   E. 精密度
_____ 6. precision   F. 探测
_____ 7. detection   G. 准确度
_____ 8. factor   H. 能量

Ⅱ. Reading.

Automation

Automation or automatic control is the use of various control systems for operating equipment. One of the primary functions of automation is to standardize repetitive production operations **so as to**[1] upgrade production rates. To do this, automation techniques most often replace inconsistent manual functions with consistent machine functions. This is where automation can and save labors. It is also used to save energy and materials and to improve **quality accuracy and precision**[2].

The process of automating a manufacturing process may **take many forms**[3]. It perhaps involves a simple mechanical device that operates part of a machine, complex computer driven feedback systems.

In a true automatic system, the automation device has an ability of detection factors when the process is being accomplished and then starts control functions.

Automation requires the following factors:

1. An automated process.

2. A sensing system for making process decisions.

3. A control action system that operates on the sensed information and then provides a **process control function**[4].

Notes: 1. so as to: 以致，以便。

2. quality accuracy and precision: 质量的精准性和精密度。

3. take many forms: 采取多种形式。

4. process control function: 过程控制功能。

阅读参考译文

Ⅲ. After-reading Activities.

Task 4   Answer the questions according to the passage.

1. What is automation?

2. What are the advantages of automation?

3. In a true automatic system, does the automation device have an ability of detection factors?

4. Do you know the factors which automation requires? What are they?

5. What's a control action system?

Task 5   Fill in the blanks according to the passage.

The process of 1. _____ a manufacturing process may take many forms. It perhaps 2. _____ a simple mechanical device that operates part of a machine, 3. _____ computer driven feedback systems.

In a true automatic 4. _____, the automation device has an ability of detection 5. _____ when the process is being accomplished and then starts control functions.

Task 6 Translate the sentences into Chinese.

1. One of the primary functions of automation is to standardize repetitive production operations so as to upgrade production rates.
2. It is also used to save energy and materials and to improve quality accuracy and precision.
3. The process of automating a manufacturing process may take many forms.
4. In a true automatic system, the automation device has an ability of detection factors.
5. Automation requires the following factors.

Task 7 Complete the following sentences by choosing an appropriate answer from the box.

primary	repeat	save	true	require	process

1. Mary _____ the words all the time.
2. I was trying to _____ money to go to college.
3. His misunderstanding of language was the _____ cause of his other problems.
4. The supervisor is very familiar with the production _____.
5. The rules also _____ employers to wear safety clothes.

Task 8 Translate the following sentences into English.

1. 使用机械设备有助于提高生产率。（upgrade）

_____

2. 我们需要一些原材料。(material)

_____

3. 我们对这台机器很感兴趣。(machine)

_____

4. 每个人应该具有分析和解决问题的能力。(an ability of)

_____

5. 自动化用于很多方面。(automation)

_____

⟳ Part Three   Let's Write

Congratulation Letter

在西方，得知朋友或同事订婚、结婚、生孩子、升职都要写信祝贺。祝贺信如同一般的简短书信，可长可短，格式上无特别的要求，但书写时应做到真诚、自然、亲切动人，要言简意赅，篇幅不宜太长。

祝贺信除了向对方道贺以外，还可表达祝福、期望等。语言应热情洋溢，满怀喜悦。写作一般分三步：

（1）说明事由并表达自己忠心的祝贺。

（2）展开事件评论，赞扬收信人。

（3）再次表达良好祝愿。

Sample 1 （齐头式）

Dear Mr. Bell,

I would like to convey my warm congratulations on your appointment to the Manager of Sales Department.

My fellow colleagues and I are delighted that the many years of service you have given to your company should at last have been rewarded in this way. We all join in sending you our very best wishes for the future.

Yours sincerely,

Mary Willis

Sample 2 （缩进式）

Aug. 16, 2018

Dear Taylor,

Here's a word for your 18th birthday from your old friend. My entire family joins me in sending you sincere & kind wishes. Please accept the small gift. I'm hoping you will like it.

Sincere yours,

Jesse

Useful Expressions

1. I'm writing to congratulate you upon… 我写信来祝贺你……。

2. I offer you my warmest congratulations on your… 对于你的……我表示热烈的祝贺。

3. Congratulations on your achievements. 祝贺你取得的成就。

4. We are delighted to learn that you have passed the examination.
我们很高兴得知你通过了考试。

5. I am glad to hear that you have got your promotion.
我很高兴听说你升职了。

6. It was with great pleasure that I read of your promotion to the position of...
得知你晋升为……我非常高兴。

7. With all the hard work you have put in recent years, this promotion is all
you deserve. 正是近年来的努力工作让你的升职实至名归。

8. Please accept our most sincere congratulations and very best wishes for a
bright future. 请接受我们最诚挚的祝愿并祝前程似锦。

9. May you enjoy good health and long life. 祝你健康长寿。

10. Congratulations and all good wishes! 祝贺您，并致以良好的祝愿！

Writing Practice

A. *Fill in the blanks in the congratulation letter below with the words and
expressions given.*

deserve	congratulation	assistance
appointment	election	in view of

Dear Mr. Calvert,

I would like to offer my sincere _____ on your _____ as the
chairman of our cooperation.

Your _____ came no surprise to us _____ your unusual abilities. No
one has done more to _____ the honor. You can depend on me and my
company to give you any _____ you require in your term of office, and I wish
you every success in the future.

Yours sincerely,

Jenny Emmett

B. *Write a congratulation letter according to the given situation.*

说明：假设你是市场部经理 Mr. Johnson。你的一个商业伙伴 Mr. Hanks 最近升职为世纪贸易公司的总经理。Mr. Johnson 现在需要给 Mr. Hanks 写一封祝贺信。

时间：9 月 12 日

Words for Reference：

商业伙伴：business partner

贸易公司：trade company

Part Four　Learning More

Ⅰ. Vocabulary

表示"状态"的相关词语

■ **be at rest** 处于静止状态

If an object is at rest in a moment, is the force on it necessarily zero?

如果一物体在某一瞬间是静止的，那么它所受到的外力是否必然为零呢？

■ **be under way** 正在进行

An investigation is under way to find out how the fire happened.
一项调查正在进行，以查明这次火灾是怎样发生的。

■ **be in operation** 正在运行

The lathes are in operation well. 车床运行良好。

■ **be in poor condition/in disrepair** 状况不佳/年久失修

The elevator is in poor condition/in disrepair. Don't use it.
电梯状况不佳/年久失修，不要使用。

■ **be in danger** 处于危险状态

These green plants are in danger because of the environmental pollution.
由于环境污染，这些绿色植物处于危险中。

■ **be in progress** 正在进行；在进行中

The project is in progress. 这个项目正在进行中。

■ **be on fire** 着火

The buildings were on fire last night. 昨晚建筑物着火了。

■ **be under discussion** 正在讨论中

The size and the form of a new product are under discussion.
新产品的大小和形状正在讨论中。

Practice

Ⓐ *Read the following sentences and underline the related words that mean* "状态".

1. Although it is raining, the marathon is still in progress.

2. The bridge was in disrepair, and it was in danger.

3. This operation system has been in operation for more than 3 years.

4. Is the paper still on fire?

5. A research by Mr. Xiao was under way.

B. *Translate the five sentences above into Chinese.*

1. _____

2. _____

3. _____

4. _____

5. _____

Ⅱ. Grammar

比较级和最高级

形容词和副词均有三种形式：原级、比较级、最高级。比较级和最高级的用法如下：

（1）原级比较的用法。形容词和副词的原级也可以用于表示比较，通常用于句型 as… as…和其否定式 not so/as… as…中。

e. g. Our neighbor's dog is as old as ours.

邻居家的狗和我们家的狗年龄一样。

The food in this restaurant is not so good as before.

这家饭店的菜没有以前好吃了。

英语中有很多由这个形式构成的固定搭配，例如：

as busy as a bee 忙碌　　　　as brave as a lion 勇敢无畏

as greedy as a wolf 贪得无厌　as loud as thunder 声如雷鸣

as silly as a goose 蠢如呆鹅　　as bright as day 亮如白昼

as stupid as a donkey 笨极了　as timid as a hare 胆小如鼠

（2）比较级 + than

e. g. Her voice is sweeter than mine. 她的声音比我的甜美。

She swam faster than I do. 她比我游得快。

（3）less + 原级 + than

e. g. The book is less popular than one. 这本书没有那本受欢迎。

（4）the more... the more...

e. g. The more you eat, the fatter you will be.
你吃的越多就会越胖。

The harder I work, the more I will learn.
学习越努力，学会的就更多。

（5）more and more

e. g. My sister's French is getting better and better.
我姐姐的法语越来越好了。

Now I see more and more clearly.
现在我看得越来越清晰了。

（6）倍数的表示方法

e. g. This room is three times as big as that one.
这个房间是那个房间的三倍大。

= This room is three times the size of that one.

= This room is twice bigger than that one.

（7）形容词最高级前常有定冠词 the，副词最高级前可将其省略。

e. g. This is the best film I have ever watched.
这是我看过的最好的电影。

She runs fastest in the class.
她是班级里跑得最快的。

Practice

Fill in the blanks.

1. clean 比较级_____ 最高级_____

2. hot 比较级_____ 最高级_____

3. carefully 比较级_____ 最高级_____

4. quickly 比较级_____ 最高级_____

5. far 比较级_____ 最高级_____

6. What do you think is _____ (important), family or work?

7. He laughs _____ (well) who laughs last.

8. Shopping online is _____ (cheap) than shopping in big stores.

9. The _____ (warm) the weather, the _____ (happy) the children are.

10. Of the three of us, I sang _____ (badly).

Unit Eight Interview

 Task1   Read the sentences aloud and do exercises.

■ Can you tell me why you want to work as...? 您能告诉我您为什么想成为……吗?

e. g. Can you tell me why you want to work as a sales manager?
您能告诉我您为什么想成为销售经理吗?

■ Do you have any experience in...? 您有……方面的经验吗?

e. g. Do you have any experience in the sales field?
您有销售方面的经验吗?

■ I don't have much... experience, but... 我没有……方面的经验,但是……

e. g. I don't have much sales experience, but this is the job I intended to take when I was at university.
我没有销售方面的经验,但是我从大学时就很想从事这方面的工作。

■ I like… most, because… 我最喜欢…… 因为……

e. g. I like sales strategies most, because I like dealing with people and it's very challenging.

我最喜欢营销策略，因为我喜欢与人打交道，那很有挑战性。

■ I hope to let you know about the result… 我想在……（时候）就能够答复您。

e. g. I hope to let you know about the result within the week.

我想在本周内就能够答复您。

Exercise 1   Fill in the blanks.

1. 您能告诉我您为什么想成为人事部经理吗？

   Can you _____ me _____ you want to _____ _____ personnel manager?

2. 您有广告方面的经验吗？

   Do you _____ any _____ in the _____ field?

3. 我没有多少经验，但是我愿意学习。

   I don't have _____ _____, but I'm willing to _____.

4. 我最喜欢计算机辅助设计（CAD），因为它在许多领域都很实用。

   I like _____ _____, because it is very practical in many fields.

5. 我想在下周日之前能够答复您。

   I _____ to let you know about the result before _____ _____.

Exercise 2   Rearrange the sentences to make a dialogue.

A. I used to work as the assistant to the general manager.

B. That's good. What do you like most about your last job?

C. What position do you apply for?

D. I see. Thank you for coming to this interview. I'll let you know about the result before next week.

E. I like dealing with all those files, because I have very good computer skills.

F. I apply for the position of the assistant manager.

G. Do you have any related experience?

_____ _____ _____ _____ _____ _____ _____

Task 2   Role-play.

Words bank

application form 申请表   mechanical engineer 机械工程师   field 领域
intern 实习生   salary 薪水   opening ceremony 开幕式
industry 工业   card 名片   show 展示

Dialogue 1

Mr. Liu   : Good morning, Mr. Parson.

Mr. Parson: Good morning, Mr. Liu. Take a seat, please.

Mr. Liu   : Thank you.

Mr. Parson: Let me look at your application form. Mr. Liu, can you tell me why you want to work as a mechanical engineer?

Mr. Liu   : At first, I think, I'm interested in the job. And your company is one of the best.

Mr. Parson: Do you have any experience in the mechanism field?

Mr. Liu   : Yes. I had been an intern for three months in the MK Mechanical Company when I was at university.

Mr. Parson: What subjects do you like most at university?

Mr. Liu   : Mechanical design.

Mr. Parson：What about your hobbies?

Mr. Liu　：I like sports. I often play volleyball.

Mr. Parson：Great. Is there anything you want to ask me?

Mr. Liu　：Yes. Can you tell me something about the holidays and salary?

Mr. Parson：There are four weeks of holidays a year, excluding public holidays. And the starting salary is about 3000 RMB a month.

Mr. Liu　：I've got it.

Mr. Parson：Well, thank you very much for coming. I hope to let you know about the result within the week.

Notes：What about...? （对于）……怎么样？相当于 how about...。
I've got it. 我明白了。I've 是 I have 的缩写形式。

Dialogue 2

Andrew　：Good morning. I'm Andrew Lee. I'm glad to have the opportunity for this interview.

Hawkins：Good morning, I'm John Hawkins. Please take a seat.

Andrew　：Thank you, Mr. Hawkins.

Hawkins：First of all, I'd like to introduce the main responsibilities of this position to you. The major responsibilities are carrying out quality inspections and responding immediately to equipment breakdowns. So anyone on this position must have good eyesight and normal color vision.

Andrew　：Yes. I can say that I have perfect eyesight. Besides, I have a lot of practical knowledge and experience from my last job.

Hawkins：Good! The job may involve either preventative or emergency maintenance. So you may need to fulfill emergency call-out duties. Are you willing to work overtime?

Andrew : Yes, I am. May I ask if there is any chance to be promoted?

Hawkins : Of course. You'll have very good prospects for promotion.

Andrew : Could you tell me the salary?

Hawkins : It will be 55,000 dollars a year.

Andrew : When can I have the result for this interview?

Hawkins : Our final decision will be available at this weekend. I'll let you know as soon as possible.

Andrew : All right. Thank you for your time, Mr. Hawkins.

Notes: take a seat: 就座，也可说成 have a seat。

call-out: 应召出勤。

as soon as possible: 尽快，可缩写成 ASAP。

Task 3 Find the correct answer for Speaker A.

Speaker A	Answer
1. Shall we arrange a meeting next week? _____	a. By credit card.
2. Why are you late for the interview? _____	b. He retired last week.
3. How do you like to pay for the computer? _____	c. From our website.
4. Why isn't Bob working here now? _____	d. No problem.
5. Excuse me, where can I learn more about your company? _____	e. I missed the bus.

Task 4   Look at the pictures，and choose the correct English expression for each item.

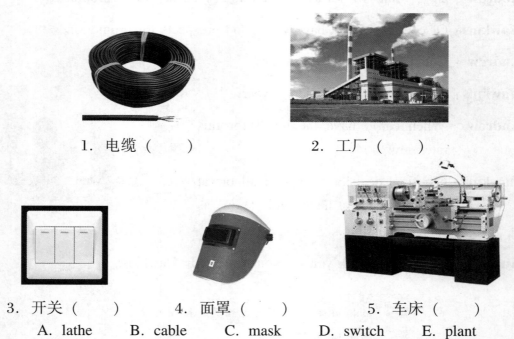

1. 电缆（   ）   2. 工厂（   ）

3. 开关（   ）   4. 面罩（   ）   5. 车床（   ）
A. lathe   B. cable   C. mask   D. switch   E. plant

 Part Two   Let's Read

Ⅰ. Pre-reading Activities.

Task 1   Read aloud the following words and learn the meanings.

discipline [ˈdɪsɪplɪn] *n.* 学科；符合行为准则的行为（或举止）	interpret [ɪnˈtɜːprɪt] *v.* 解释；诠释
architect [ˈɑːkɪˌtekt] *a.* 建筑师；缔造者	draftsperson [ˈdrɑːftspɜːs(ə)n] *n.* 绘图员
interior [ɪnˈtɪərɪə(r)] *a.* 内部的	graphic [ˈɡræfɪk] *a.* 图解的；用图表示的
thickness [ˈθɪknəs] *n.* 厚度	specification [ˌspesɪfɪˈkeɪʃ(ə)n] *n.* 规格；说明书
notation [nəʊˈteɪʃ(ə)n] *n.* 记号；标记法	dimension [daɪˈmenʃ(ə)n] *n.* 方面；维

Task 2   Write the words according to the English explanation
（Hints：the words from task 1.）

1. _____ a skilled worker who draws plans of buildings or machines.
2. _____ someone who creates plans to be used in making something（such as buildings）.
3. _____ inside.
4. _____ a set of written symbols.
5. _____ a detailed description.
6. _____ give an interpretation or explanation to.

Task 3   Match the English words with the correct item.

_____1. technical drawing   A. 专业人员
_____2. design engineer   B. 技术制图
_____3. professional   C. 制图员
_____4. cartographer   D. 几何
_____5. view projection   E. 目的
_____6. geometry   F. 视图投影
_____7. technician   G. 设计工程师
_____8. purpose   H. 技术员

II. **Reading.**

Technical Drawing

Technical drawing is also called drafting. **This academic discipline is created by**[1] architects, interior designers, drafters, design engineers, and related professionals. It is used to create **standardized technical drawings**[2] which include standards and conventions for

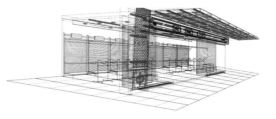

layout, line thickness, text size, symbols, view projections, descriptive geometry, dimensioning, and notation.

A person who does drafting is known as a drafter. In some areas this person may be called a drafting technician, draftsperson, or draughtsman. This person creates technical drawings which are a form of specialized graphic communication. The difference between a technical drawing and a common drawing is how to interpret. A common drawing can have many purposes and meanings, while a technical drawing can indicate all needed specifications concisely and transform an idea into physical form.

Drafting is the general term of mechanical or engineering drawings and is the **sub-discipline**[3] of industrial arts. It can be done in **two dimensions（"2D"）and three dimensions（"3D"）**[4].

阅读参考译文

Notes: 1. This academic discipline is created by... 是被动语态，译为"由……创建这门学术学科"，介词 by 表示"由"。
2. standardized technical drawings：标准技术图样。
3. sub-discipline：分支学科。
4. two dimensions（"2D"）and three dimensions（"3D"）：二维和三维。

Ⅲ. After-reading Activities.

Task 4　Answer the questions according to the passage.

1. Which subject is created by architects, interior designers, drafters, design engineers, and related professionals?
2. What do standardized technical drawings include?
3. What does a drafter do?
4. What is the difference between a technical drawing and a common drawing?
5. What is drafting?

Task 5   Fill in the blanks according to the passage.

A person who does drafting is known as a 1. _____. In some areas this person may be called a drafting 2. _____, draftsperson, or draughtsman. This person creates technical drawings which are a form of specialized graphic communication. The 3. _____ between a technical drawing and a common drawing is how to interpret. A common drawing can have many purposes and meanings, 4. _____ a technical drawing can indicate all needed specifications concisely and clearly and transform an idea 5. _____ physical form.

Task 6   Translate the sentences into Chinese.

1. Technical drawing is also called drafting.
2. In some areas this person may be called a drafting technician, draftsperson, or draughtsman.
3. The difference between a technical drawing and a common drawing is how to interpret.
4. A technical drawing can indicate all needed specifications concisely and transform an idea into physical form.
5. Drafting is the general term of mechanical or engineering drawings and is the sub-discipline of industrial arts.

Task 7   Complete the following sentences by choosing an appropriate answer from the box.

| academic | layout | size | a form of | interpret | concisely |

1. This map shows the _____ of the park.
2. What's the _____ of the red shirt?
3. The speaker will attend an _____ meeting in Beijing next week.
4. Forgetfulness is _____ freedom.

5. The language points are summed up _____ in the book.

Task 8 Translate the following sentences into English.

1. 这门学科由制图员、设计工程师和相关专业人士创建。（create）

_____

2. 在某些领域这种人可能被称作机械工程师。（call）

_____

3. 这是中国和加拿大的不同之处。（difference）

_____

4. 吉姆是美国一个很常见的名字。（common）

_____

5. 我们如何转变文件类型？（transform）

_____

Part Three   Let's Write

Application Letter

　　求职信在找工作的过程中是非常重要的。一封能够体现个人才智的求职信，能够帮助你的求职之路更加顺利。好的求职信会给招聘方留下深刻的印象而使求职者获得面试机会。求职信的语言要简洁明了，用词妥帖，避免使用专业术语和复杂句子；语言客观实际，自信而不浮夸；语法准确无误，拼写正确。务必注意对方姓名、公司名称等重要信息不能出错，而且格式要正式、排版大方美观，内容最好在一页之内。

　　求职信的主要内容一般包括：

　　（1）介绍信息来源及求职心愿。

（2）说明个人基本情况，应届毕业生着重介绍所学课程和兼职工作经验，具有工作经验的人要突出自己的工作经验和业绩等。

（3）列出其他优势和特长。

（4）表达感谢，提出希望得到面试机会并留下联系方式。

求职信主要由七部分构成：

（1）信头 heading（sender's address）。

（2）信内地址 inside address（receiver's address）。

（3）称呼 salutation。

（4）正文 body。

（5）结束语 complimentary close。

（6）签名 signature。

（7）附件 enclosure。

格式如下：（齐头式格式）

Heading (寄信人的地址和日期)

P. O. Box 64

Nanjing University

Nanjing，China 100084

April 7，2018

Inside Address(收信人地址)

Prit Co.，Ltd.

1803 International Office Building

65 Taoxi Road

Hangzhou，325000

Salutation (称呼语)

Dear Sir/Madam,

Body (正文)

I am applying for the post of Marketing Representative advertised in China Daily，dated 27 March 2018.

I graduated from Nanjing University in June with a bachelor degree in Business

Management. At the college I served as secretary of the Student's Union and was responsible for the operation of a book shop for students. As your company is expanding to the food market, my experience would be a good choice for you.

Thank you for your time to read the letter. I would welcome an interview at any time convenient for you. You can call me at 13734598765 or send an email to me at mayjincheng9601@163. com.

Thank you very much.

Complimentary Close
结束礼词

Yours faithfully,

Signature (签名)

May Jincheng

Enclosure(附件)

Enclosed— resume

Sample （此例文格式为缩进式）

School of Mathematical Sciences
Peking University
Bejing, China 100084
June 15, 2018

Personnel Department
Lexco Trade Co.
No.36 Zhonghuan Street
Guangzhou 510014

Dear Mr. Brown,

I noted with interest about your advertisement in today's *Bejing Daily*.

The position you described sounds exactly like the kind of job I am seeking for. You will see from the resume that I am a college graduate with great enthusiasm and ambition. I can cover all the abilities required in the advertisement because I have ever taken an internship as an analyst in a similar company.

I will graduate from the college in this coming July. Although I am a graduate without much experience, I've spent lots of time developing my communication skills and teamwork spirits.

If you are interested in my application, I would be available for interview at any time.

Yours sincerely.
Wu Yuesha

Enclosure: resume

Useful Expressions

1. I'm writing to apply for the position of... 我写信申请……职位

2. I would like to apply for the position of... 我想申请……的职位

3. I am a graduate of Dongfang Technical College.
   我是东方技术学院的毕业生。

4. I believe these skills will enable me to work well in your team.
   我相信我拥有的这些技能能让我在你的团队中工作的很好。

5. My working experience at... company has improved my... and... ability.
   我在……公司的工作经历让我提高了我的……能力。

6. I'm experienced in operating modern office equipments.
   本人对现代办公设备操作熟练。

7. I have excellent English writing and reading ability.
   我拥有优秀的英语读写能力。

8. If you need any additional information, or if you have any questions, please contact me at...
   如果您需要更多信息或有什么问题，请致电……

9. I am available for interview at any time. 我随时可以参加面试。

10. Thank you for your consideration and look forward to hearing from you.
    感谢您的关注，敬盼佳音。

Writing Practice

A. *Translate the following sentences.*

1. 我似乎符合贵公司广告中所提出的要求。

2. 我想应聘销售经理一职。

3. 我具有多年从业经验，熟悉该岗位的各项工作，是这个岗位的最佳人选。

4. 我性格开朗，容易相处，具有良好的团队协作精神。

5. 我希望您能给我一个面试的机会，让您更进一步了解我的优点。

B. *Write an application letter according to the information.*

说明：假定你是张天华，根据以下内容写一封求职信。

写信日期：2018 年 5 月 6 日

信函内容：Lex 电子公司在《北京日报》上刊登广告招聘一名行政秘书。张天华是北京电子信息学院的一名即将毕业的学生，专业是电子信息，选修过企业管理、行政秘书等课程。

Words for reference：

行政秘书：administrative secretary

电子信息：electronic information

企业管理：business management

⟳ Part Four   Learning More

Ⅰ. Vocabulary

表示"引起"的相关词语

■ **cause** *v.* 引起；产生

Smoke can cause your health problem.

吸烟能引发你的健康问题。

■ **induce** *v.* 引起，导致

Surgery may induce a heart attack. 手术可能导致心脏病。

■ **be due to...** 是由……所引起的

The expansion of a solid is due to the increase in space between the molecules.

固体的膨胀是由分子之间的间隔扩大所致。

■ **generate** *v.* 产生；引起

The application of high technology generates profit for BOM Company.

高科技的应用为 BOM 公司产生利润。

■ **give rise to...** 引起

Poor maintenance gives rise to wear of equipment.

维护不良会引起设备磨损。

■ **produce** *v.* 产生；引起

Impurities produce harmful effects. 杂质会引起有害的影响。

Practice

Ⓐ *Read the following sentences and underline the related words that mean* "引起".

1. Operation of the machine gives rise to vibrations.

2. High temperature induces combustion.

3. A shortage of electricity is due to a shortage of coal.

4. How can the device generate airflow?

5. Rapid expansion of a gas causes its temperature to fall.

B. *Translate the five sentences above into Chinese.*

1. _____

2. _____

3. _____

4. _____

5. _____

Ⅱ. Grammar

非真实条件句

if 引导的条件句中，如果所表示的意思可能实现，这种条件句叫作真实条件句；如果所表达的是一种假想的情况或主观愿望，这种条件句就叫作非真实条件句，也叫虚拟条件句，需要使用虚拟语气与真实条件句加以区分。

非真实条件句根据动作发生的时间分为现在、过去和将来三种情况。

（1）表示现在的情况

构成形式：if 条件从句，动词用 did（be 动词用 were），主句用 would do。

e. g. If I were a boy, I would join in the army.

如果我是男孩，我就去参军。

If I lived on that street, I would go to the park every morning.

如果我住在那条街上，我每天都会去公园。

（2）表示过去的情况

构成形式：if 条件从句用 had done，主句用 would have done。

e. g. If I had left home earlier, I would't have missed the train.

如果我早些出门，我就不会错过火车了。

If you had taken my advice, you would have winned the competition.
如果你听取我的建议，你就可能赢得比赛。

（3）表示将来的情况

构成形式：if 条件从句用 did/were to do/should do，主句用 would do。

e. g.   If Tom knew her number, he would call her this afternoon.
如果汤姆知道她的电话号码，他今天下午就会给她打电话。

If I were to do the job, I would finish it ahead of schedule.
如果我来做这项工作，我会提前完成。

If I should be free tomorrow, I would come to pick you up in the morning. 如果我明天没事，我上午会来接你。

在这三种情况中，主句中的助动词 would 也可替换成 could、might、should。

e. g.   If you tried very hard, you might succeed.
如果你非常努力，你可能会成功。

条件从句中 if 有时可以省略，这时从句部分需要部分倒装：

e. g.   Were I you, I wouldn't take the job.
如果我是你，我就不会接受这份工作。

Had the boy worked harder, he wouldn't have failed the exam.
如果男孩学习再努力些，他就不会考试不及格。

Practice

Fill in the blanks.

1. If I _____ (see) Luke, I would ask him.

2. If I _____ (know) you were coming, I would have picked you up.

3. If I _____ (tell) her yesterday, she wouldn't have made that mistake.

4. If the earth _____ (stop) spinning, everything on it would fly off.

5. _____ we _____ (leave) earlier on Monday, we would have arrived in Tokyo.

6. _____ (be) I you, I wouldn't agree.

7. If it _____ (not be) for your help, we couldn't get out of the trouble so quickly.

8. If I had sufficient food, I _____ (give) some to that poor man.

9. If you had arrived earlier, you _____ (see) Aunt Susan.

10. If anything went wrong, he _____ (let) me know.

Vocabulary

A

accomplish [əˈkʌmplɪʃ] v. 完成；达到（目的）

activate [ˈæktɪˌveɪt] v. 使活动；使开始作用

application [ˌæpliˈkeɪʃ(ə)n] n. 应用

architect [ˈɑːkɪˌtekt] a. 建筑师；缔造者

artificial [ˌɑːtɪˈfɪʃ(ə)l] a. 人造的；人工的

ascribe [əˈskraɪb] v. 把……归于；认为……具有

assembly [əˈsembli] n. 装配；集合

assistance [əˈsɪstəns] n. 帮助；援助

automation [ˌɔːtəˈmeɪʃ(ə)n] n. 自动化

available [əˈveɪləb(ə)l] a. 可获得的；可用的

B

blower [ˈbləʊə] n. 鼓风机

C

catalog [ˈkætəlɒg] n. 目录，目录册

chamber [ˈtʃeɪmbər] n. 议会；议院

clamp [klæmp] v. 夹紧，夹住

cover [ˈkʌvə(r)] n. 覆盖物

D

danger [ˈdeɪndʒə(r)] n. 危险

device [dɪˈvaɪs] n. 装置，设备

derive [dɪˈraɪv] v. 从……衍生；起源于

dimension [daɪˈmenʃ(ə)n] n. 方面；维

discipline [ˈdɪsɪplɪn] n. 学科；符合行为准则的行为（或举止）

draftsperson [ˈdrɑːftspɜːs(ə)n] n. 绘图员

drive [draɪv] n. 驱动器；驱动力

E

edit [ˈedɪt] v. 编辑

electric [ɪˈlektrɪk] a. 电的；电动的；发电的

enclose [ɪnˈkləʊz] v. （随信）附上

essentially [ɪˈsenʃ(ə)li] ad. 本质上，根本上

execution [ˌeksɪˈkjuː(ə)n] n. 实行，执行

explosion [ɪkˈspləʊʒ(ə)n] n. 爆炸

exporter [ˈekspɔːtə(r)] n. 出口商

F

faithfully [ˈfeɪθfəli] ad. 忠实地；诚心诚意地

fan [fæn] n. 风扇

feedback [ˈfiːdˌbæk] n. 反馈；反应

field [fiːld] n. 领域；范围

G

goods [gʊdz] n. 商品

graphic [ˈgræfɪk] a. 图解的，用图表示的

H

hazard [ˈhæzəd] n. 危险；冒险

hook [hʊk] n. 钩，铁钩

I

inconsistent [ɪnkənˈsɪst(ə)nt] a. 不一致的

industrial [ɪnˈdʌstrɪəl] a. 工业的；产业的

intelligent [ɪnˈtelɪdʒ(ə)nt] a. 智能的

interior [ɪnˈtɪəriə] a. 内部的

interpret [ɪnˈtɜːprɪt] v. 解释；诠释

invent [ɪnˈvent] v. 发明

item ['aɪtəm] *n.* 条款；项目

K

keyboard ['kiːbɔːd] *n.* 键盘

L

ladder ['lædə(r)] *n.* 梯状物

M

machinery [mə'ʃiːn(ə)ri] *n.* 机械；机器

manipulate [mə'nɪpjuˌleɪt] *v.* 操作，处理

manual ['mænjʊəl] *n.* 手册；指南

microprocessor [ˌmaɪkrəʊ'prəʊsesə(r)] *n.* 微处理器

minimize ['mɪnɪˌmaɪz] *v.* 把……减至最低数量[程度]

mission ['mɪʃ(ə)n] *n.* 使命；任务

monitor ['mɒnɪtə(r)] *n.* 显示器；监视器；监控器

motion ['məʊʃ(ə)n] *n.* 运动；动机

motor ['məʊtə(r)] *n.* 电动机；发动机

N

normally ['nɔːm(ə)li] *ad.* 通常地；一般地

notation [nəʊ'teɪʃ(ə)n] *n.* 记号；标记法

O

operate ['ɒpəreɪt] *v.* 操作；经营

output ['aʊtˌpʊt] *n.* 输出

overhang [ˌəʊvə'hæŋ] *v.* 突出；伸出

P

perception [pə'sepʃ(ə)n] *n.* 知觉；觉察

portable ['pɔːtəb(ə)l] *a.* 手提的，便携式的

possess [pə'zes] *v.* 拥有

primary ['praɪməri] *a.* 首要的；主要的

process ['prəʊses] *n.* 过程

production [prə'dʌkʃ(ə)n] *n.* 生产；产品

protective [prə'tektɪv] *a.* 防护的

prove [pruːv] *v.* 证明，证实

pump [pʌmp] *n.* 泵

Q

quality ['kwɒlɪti] *n.* 质量

R

rectifier ['rektɪfaɪə] *n.* 整流器

reliable [rɪ'laɪəb(ə)l] *a.* 可靠的；真实可信的

replace [rɪ'pleɪs] *v.* 取代，代替

reprogrammable [rɪ'prəʊgræməbl] *a.* 可编程序的

S

satisfy ['sætɪsˌfai] *v.* 令人满意

sensing ['sensɪŋ] *n.* 感觉；指向

sensory ['sensəri] *a.* 感官的；传递感觉的

slow down （使）慢下来

specification [ˌspesɪfɪ'keɪʃ(ə)n] *n.* 规格；说明书

standardize ['stændərdaɪz] *v.* 使标准化

switch [swɪtʃ] *n.* 开关

symbol ['sɪmb(ə)l] *n.* 标志；符号

T

technique [tek'nik] *n.* 技术；技艺

thickness ['θɪknəs] *n.* 厚度

V

valve [vælv] *n.* 阀；真空管

vibration [vaɪ'breɪʃ(ə)n] *n.* 摆动；振动

volume ['vɒljuːm] *n.* 量；体积

W

workshop ['wɜːkʃɒp] *n.* 车间；工场

参 考 文 献

［1］田芳. 新机电英语 ［M］. 北京：机械工业出版社，2014.

［2］姜少杰，王永鼎. 机电工程专业英语 ［M］. 北京：机械工业出版社，2009.

［3］朱林，杨春杰. 机电工程专业英语 ［M］. 北京：北京大学出版社，2010.

［4］杨晓辉，刘丽红，许晶. 机电专业英语 ［M］. 北京：北京理工大学出版社，2013.

［5］戴文进，章卫国. 自动化专业英语 ［M］. 武汉：武汉理工大学出版社，2006.

［6］《机电英语》教材编写组. 机电英语 ［M］. 北京：高等教育出版社，2010.

［7］胡庭山. 机电英语 ［M］. 北京：外语教学与研究出版社，2009.

［8］张道真. 实用英语语法 ［M］. 北京：外语教学与研究出版社，2009.

［9］耿小辉，昂秀，外语教学研究组. 日常交际英语口语900句 ［M］. 北京：中国出版传媒股份有限公司，2014.

［10］步雅芸. 商务英语写作 ［M］. 北京：北京大学出版社，2010.

［11］张大为，贺春霞. 高职高专英语语法 ［M］. 北京：北京理工大学出版社，2011.

［12］袁懋梓. 大学英语语法 ［M］. 北京：外语教学与研究出版社，2008.

［13］罗明星，邱世凤. 大学英语实用写作 ［M］. 成都：西南交通大学出版社，2007.

［14］陈敏，刘东霞. A级综合训练教程 ［M］. 长春：东北师范大学出版社，2014.

［15］刘恺. 英语应用能力考试备考试题详解 ［M］. 北京：旅游教育出版社，2011.

［16］徐飞跃，郭丽丽. 高等学校英语应用能力考试B级教程精编 ［M］. 上海：上海交通大学出版社，2012.

［17］耿静先. 商务英语翻译教程 ［M］. 北京：中国水利水电出版社，2010.